SpringerBriefs in Applied Sciences and Technology

Series editor

Janusz Kacprzyk, Polish Academy of Sciences, Systems Research Institute, Warsaw, Poland

W0193179

SpringerBriefs present concise summaries of cutting-edge research and practical applications across a wide spectrum of fields. Featuring compact volumes of 50–125 pages, the series covers a range of content from professional to academic.

Typical publications can be:

- A timely report of state-of-the art methods
- An introduction to or a manual for the application of mathematical or computer techniques
- A bridge between new research results, as published in journal articles
- A snapshot of a hot or emerging topic
- An in-depth case study
- A presentation of core concepts that students must understand in order to make independent contributions

SpringerBriefs are characterized by fast, global electronic dissemination, standard publishing contracts, standardized manuscript preparation and formatting guidelines, and expedited production schedules.

On the one hand, **SpringerBriefs in Applied Sciences and Technology** are devoted to the publication of fundamentals and applications within the different classical engineering disciplines as well as in interdisciplinary fields that recently emerged between these areas. On the other hand, as the boundary separating fundamental research and applied technology is more and more dissolving, this series is particularly open to trans-disciplinary topics between fundamental science and engineering.

Indexed by EI-Compendex and Springerlink.

More information about this series at http://www.springer.com/series/8884

Abdul Jabbar

Sustainable Jute-Based Composite Materials

Mechanical and Thermomechanical Behaviour

 Springer

Abdul Jabbar
Department of Material Engineering,
  Faculty of Textile Engineering
Technical University of Liberec
Liberec
Czech Republic

and

Faculty of Engineering and Technology
National Textile University
Faisalabad
Pakistan

ISSN 2191-530X   ISSN 2191-5318   (electronic)
SpringerBriefs in Applied Sciences and Technology
ISBN 978-3-319-65456-0   ISBN 978-3-319-65457-7   (eBook)
DOI 10.1007/978-3-319-65457-7

Library of Congress Control Number: 2017948640

Printed on acid-free paper

This Springer imprint is published by Springer Nature
The registered company is Springer International Publishing AG
The registered company address is: Gewerbestrasse 11, 6330 Cham, Switzerland

Foreword

Natural fiber reinforced polymer composites (NFPC) have gained considerable attention in the recent years due to their environment and economic benefits and low energy demand in production. The use of cellulosic stiff reinforcing fillers in polymer composites have also attracted significant interests of material scientists. Waste of natural fibers, produced in the textile industry during mechanical processing, offers a cheaper source of availability for the preparation of these stiff cellulose fibrils/fillers. The poor adhesion between the fiber and polymer matrix is also considered a major drawback in the use of natural fiber composites. Therefore, surface modification of natural fibers is necessary before using them as a reinforcement in composites.

The work presented here is based on the Ph.D. thesis of Mr. Abdul Jabbar on sustainable composite materials based on jute and highlights the economical use of textile waste to improve composite properties and addresses the issue of compability between reinforcement and matrix. The book is organized into seven chapters. Chapter 1 provides the general overview on the use of natural fibers in composites industry and objectives of research. Chapter 2 reviews the published literature on natural fibers and their polymer composites. Chapter 3 describes the details of experimental setup, whereas Chaps. 4, 5, and 6 present the analysis of results. Finally, conclusions are summarized in Chap. 7, including recommendations for future work. The text is quite comprehensive and fulfills the objectives outlined. As my former student, he has done all his work quite systematically with specific objectives under my guidance, organized, and analyzed data scientifically, and discussion of the results is quite logical with citations of previous work where necessary.

I always found him hardworking, consistent, and committed to his work. His publication activities show that he is emerged as a good researcher during his stay in TUL. I hope that this publication will be helpful to the wider research community in this field.

Liberec, Czech Republic   Prof. Jiří Militký
March 2017

Acknowledgements

Educating the mind without educating the heart is no education at all

—Aristotle

Foremost, I would like to express the deepest appreciation and respect to my supervisor, Prof. Ing. Jiří Militký, CSc. EURING, for his inspiration, guidance, and giving me an opportunity to work under his kind supervision. He has been encouraging and supportive throughout my entire time at Technical University of Liberec. I am indebted to Ing. Jana Drašarová, Ph.D. (Dean of Faculty of Textile Engineering), Ing. Gabriela Krupincová, Ph.D. (Vice Dean for Science and Research), and Ing. Pavla Tešinová Ph.D. (Vice Dean for International Affairs) for financial support during the whole period of my research. Further, I would like to thank prof. Ing. Jakub Wiener, Ph.D. for his valuable suggestions during my work. I am also thankful to Dr. Vijay Baheti for his interest in my work and advice time to time.

I express my gratitude to the management of National Textile University, Faisalabad, Pakistan, for believing in me and granting me the study leave.

I am especially grateful to Ing. Petr Hornik, Ph.D. who helped me for creep and dynamic mechanical testing of composites, Ing. Martin Stuchlik of the Institute for Nanomaterials, Advanced Technologies and Innovation TUL for providing FTIR spectra, and Ing. Jana Grabmüllerová for SEM study.

I would also like to thank all colleagues in the Faculty of Textile Engineering, especially Ing. Hana Cesarová Netolická, Kateřina Štruplová, Ing. Hana Musilová, Bohumila Keilová, and Martina Čimburová for their regular help and support. Finally, especial thanks to all friends in the doctoral study program of Faculty of Textile Engineering (especially Muhammad Usman Javed) for their support and fruitful time that we spent together in Liberec.

Last but not least, where I am today is only because of the prayers of my family, especially my father and continuous support and inspiration from my wife and kids.

Abdul Jabbar

Contents

Chapter 1
Introduction

Abstract This chapter briefly describes the interest in the use of natural fibers as reinforcement in polymer composites. It also reviews the advantages to replace synthetic counterparts in certain applications. It goes further with the aims and specific objectives of present research.

Keywords Natural fibers · Biocomposites · Lignocellulosic

1.1 Background

The impact of global climatic change is quite visible in the recent years due to increase in greenhouse gas emissions. Synthetic fibers, whose main feedstock is petroleum, are being widely used in polymer composites because of their high strength and stiffness. However, these fibers have serious drawbacks in terms of their non-biodegradability, toxicity, initial processing costs, recyclability, energy consumption, machine abrasion, and health hazards, etc (Kabir et al. 2012). Therefore, the increasing environmental awareness and international legislations to reduce greenhouse gas emissions have forced the material scientists and researchers to shift their attention from synthetic fibers to natural/renewable fibers. Natural fibers are now increasingly used as reinforcement in biocomposites because of many advantages such as cost-effectiveness, lightweight, easy to process, renewable, recyclable, available in huge quantities, low fossil-fueled energy requirements and the most importantly their high specific strength-to-weight ratio (Ku et al. 2011). This is of distinctive importance especially in interior transportation applications as it leads to reduction of vehicle weight for higher fuel efficiency, reduction in cost, and energy saving. Thus, natural fibers are considered promising candidates for replacing conventional synthetic reinforcing fibers in composites for semistructural and structural applications (Dányádi et al. 2010). Biocomposites are the composites in which natural fibers are reinforced with either biodegradable or non-biodegradable matrices (Lu and Oza 2013).

A. Jabbar, *Sustainable Jute-Based Composite Materials*, SpringerBriefs in Applied Sciences and Technology, DOI 10.1007/978-3-319-65457-7_1

Plant-based natural fibers are most commonly used lignocellulosic fibers in composite applications (Dicker et al. 2014). These fibers are derived from various parts of plants such as stems, leaves, and seeds. The fibers derived from stem (bast fibers) such as jute, flax, hemp, and kenaf are more commonly used for reinforcement in composites due to their high tensile strength and high cellulose content (Li et al. 2007). Lignocellulosic fibers mainly consist of cellulose microfibrils in an amorphous matrix containing lignin and hemicellulose. The percentage composition of each component varies for different fibers. However, cellulose is the major framework component in these fibers having 60–80% weightage and is responsible for providing strength, stiffness, and structural stability to the fiber (Saheb and Jog 1999). Among lignocellulosic fibers, jute is an abundant natural fiber used as a reinforcement in biocomposites (Corrales et al. 2007) and occupies the second place in terms of world production levels of cellulosic fibers after cotton (Cai et al. 2000).

The properties and aspect ratio of fibers and interfacial interaction between fibers and matrix govern the properties of composites. Good interfacial adhesion between fiber and polymer plays an important role in the transfer of stress from matrix to fiber and thus contributes to better performance of composite. Despite parallel advantages of lignocellulosic fibers, there is some drawback regarding their behavior in polymer matrix apart from their performance and processing limitations. These fibers have poor compatibility with several polymer matrices. Weak fiber/matrix interface reduces the reinforcing efficiency of fibers due to less stress transfer from the matrix to the fiber resulting in a poor performance of composite (Hong et al. 2008). To enhance the compability between fiber and matrix, different physical (Mukhopadhyay and Fangueiro 2009), chemical (Li et al. 2007), and biological (Li and Pickering 2008) treatments are used by researchers for fiber surface modification. However, the use of some novel and environment-friendly methods such as laser, ozone, and plasma is less common. Moreover, stiff micro/nanocellulose fillers as reinforcing element in polymer matrices are also considered promising candidates in the improvement of interface interaction and hence the performance of composites.

1.2 Objectives of Research

The overall objectives of this research are to investigate the effect of addition of stiff cellulose microfibrils, nanocellulose extraction from jute waste and its coating over woven jute reinforcement, some novel environment-friendly fiber treatment methods, and characterization of the bulk properties such as mechanical, creep, and dynamic mechanical properties of composites. Jute has been selected as the reinforcing fiber due to its good mechanical properties along with other advantages such as very low cost, easy availability, and renewability. Jute waste obtained from a jute processing mill is used as a low-cost source for producing cellulose microfillers and nanocellulose extraction. The green epoxy has been chosen as a matrix because of its high biobased contents and low petroleum-derived contents. The specific objectives are as follows:

1. To investigate the incorporation of pulverized micro jute fibrils prepared from jute waste on the mechanical and dynamic mechanical properties of alkali-treated woven jute/green epoxy composites.
2. To characterize the mechanical and dynamic mechanical properties of green epoxy composites reinforced with nanocellulose-coated jute fabric.
3. To investigate the influence of some novel treatment methods such as CO_2 pulsed infrared laser, ozone, enzyme and plasma on the creep, and dynamic and mechanical properties of woven jute/green epoxy composites.
4. To model the short-term creep data of composites using four parameters (Burger's) model and to predict the long-term creep performance based on experimental data using three different creep models, i.e., Burger's model, Findley's power law model, and two-parameter power law model.

References

Cai, Y., David, S., & Pailthorpe, M. (2000). Dyeing of jute and jute/cotton blend fabrics with 2:1 pre-metallised dyes. *Dyes and Pigments, 45*(2), 161–168.

Corrales, F., Vilaseca, F., Llop, M., Girones, J., Mendez, J., & Mutje, P. (2007). Chemical modification of jute fibers for the production of green-composites. *Journal of Hazardous Materials, 144*(3), 730–735.

Dányádi, L., Móczó, J., & Pukánszky, B. (2010). Effect of various surface modifications of wood flour on the properties of PP/wood composites. *Composites Part A Applied Science and Manufacturing, 41*(2), 199–206.

Dicker, M. P., Duckworth, P. F., Baker, A. B., Francois, G., Hazzard, M. K., & Weaver, P. M. (2014). Green composites: A review of material attributes and complementary applications. *Composites Part A Applied Science and Manufacturing, 56,* 280–289.

Hong, C., Hwang, I., Kim, N., Park, D., Hwang, B., & Nah, C. (2008). Mechanical properties of silanized jute—polypropylene composites. *Journal of Industrial and Engineering Chemistry, 14*(1), 71–76.

Kabir, M., Wang, H., Lau, K., & Cardona, F. (2012). Chemical treatments on plant-based natural fibre reinforced polymer composites: An overview. *Composites Part B Engineering, 43*(7), 2883–2892.

Ku, H., Wang, H., Pattarachaiyakoop, N., & Trada, M. (2011). A review on the tensile properties of natural fiber reinforced polymer composites. *Composites Part B Engineering, 42*(4), 856–873.

Li, X., Tabil, L. G., & Panigrahi, S. (2007). Chemical treatments of natural fiber for use in natural fiber-reinforced composites: A review. *Journal of Polymers and the Environment, 15*(1), 25–33.

Li, Y., & Pickering, K. L. (2008). Hemp fibre reinforced composites using chelator and enzyme treatments. *Composites science and technology, 68*(15), 3293–3298.

Lu, N., & Oza, S. (2013). Thermal stability and thermo-mechanical properties of hemp-high density polyethylene composites: Effect of two different chemical modifications. *Composites Part B Engineering, 44*(1), 484–490.

Mukhopadhyay, S., & Fangueiro, R. (2009). Physical modification of natural fibers and thermoplastic films for composites—A review. *Journal of Thermoplastic Composite Materials, 22*(2), 135–162.

Saheb, D. N., & Jog, J. (1999). Natural fiber polymer composites: a review. *Advances in Polymer Technology, 18*(4), 351–363.

Chapter 2
Literature Review

Abstract Due to specific advantages over synthetic counterparts, plant-based natural fibers are considered promising candidates for reinforcement in polymer composites for certain applications. This chapter deals with the classification, structure, and chemical composition of plant-based natural fibers. Some important aspects of jute fiber and its use as reinforcement in polymer composites are discussed. The use of cellulose nano fibrils as a filler in polymer composites is overviewed. Different chemical, physical, and biological methods for surface modification of natural fibers are discussed in detail. At the end, the importance of creep property in polymer composites and some important creep models have also been dealt with.

Keywords Plant fibers · Cellulose nano-fibers · Natural fiber composites · Surface modification

2.1 Natural Fiber Composites

The use of natural fibers as reinforcement in polymer composites is constantly increasing. Currently, the use of natural fiber composites is limited to interior and non-structural applications due to their poor moisture resistance and low mechanical properties (Dittenber and GangaRao 2012). These are being used in architectural, furniture, and automotive industries (Araujo et al. 2008). However, the research is underway to expand their applications by encountering the challenges associated with the use of natural fibers in polymer composites. A brief understanding of the nature, classification, and composition of natural fibers is presented in the following sections.

2.2 Natural Fibers and Their Classification

Fibers can be classified into two groups on the basis of production of fibrous polymers and production of fibers: natural fibers and man-made fibers. Natural fibers are those which occur in nature in the form of fibers, whereas man-made

© The Author(s) 2017   5
A. Jabbar, *Sustainable Jute-Based Composite Materials*, SpringerBriefs in Applied Sciences and Technology, DOI 10.1007/978-3-319-65457-7_2

Fig. 2.1 Classification of plant-based natural fibers used as reinforcement in composites (Chandramohan and Marimuthu 2011)

fibers are those which are produced by spinning from polymer prepared by humans (synthetic fibers) or occurring naturally (chemical fibers). Natural fibers are further classified according to the nature of their source into vegetable/plant, animal, and mineral fibers. Plant-based natural fibers are mostly used as a reinforcing element in composites which are further classified as shown in Fig. 2.1, on the basis of their origin. Synthetic fibers whose feedstock are fossil fuel are the leading causes of environmental degradation due to the toxicity of the emitted fumes and non-biodegradability, whereas natural fibers have advantages such as biodegradability, renewability, low cost, and non-toxicity.

2.3   Chemical Composition and Structure of Plant-Based Natural Fibers

The elementary plant fiber is a single cell having length ranging from 1 to 50 mm and diameter from 10 to 50 μm. Plant fibers are like microscopic tubes, i.e., cell walls surrounding the central lumen. The lumen is usually responsible for the water uptake behavior of plant fibers (Tsoumis 1991). The fiber contains several cell walls. These cell walls are formed from oriented semicrystalline reinforcing cellulose microfibrils embedded in a matrix of pectines, hemicellulose, and lignin of varying composition. Such microfibrils have typically a diameter in the range of 10–30 nm and are made up of 30–100 cellulose molecules in extended chain conformation and provide mechanical strength to the fiber. The typical arrangement of fibrils, microfibrils, and cellulose in the cell walls of a plant fiber is shown in Fig. 2.2.

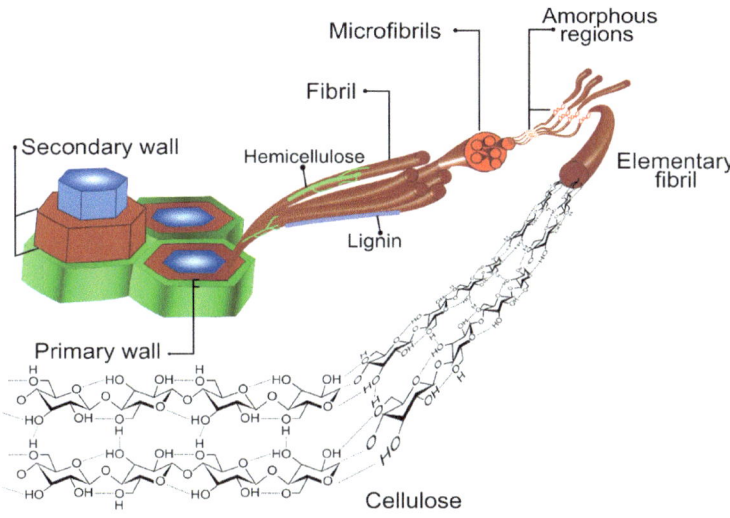

Fig. 2.2 Arrangement of fibrils, microfibrils, and cellulose in the plant cell wall (Rojas et al. 2015)

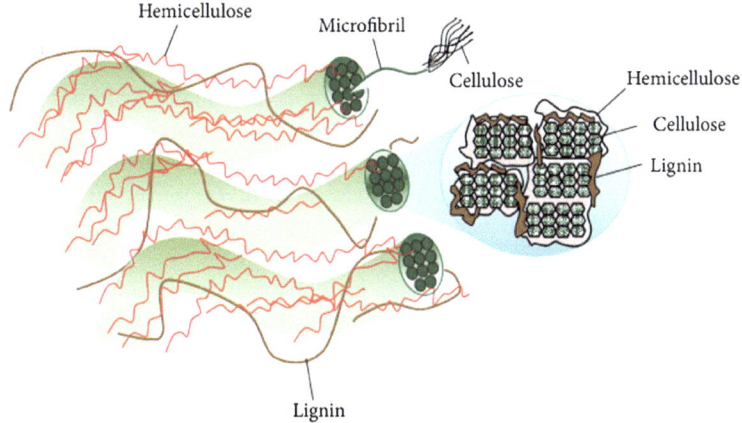

Fig. 2.3 Cellulose fibrils embedded in a matrix of hemicellulose and lignin (Lee et al. 2014)

The typical cell wall structure of plant fiber is shown in Fig. 2.3. The cellulose is hydrogen bonded to hemicellulose molecules of the matrix phase in a cell wall. Hemicelluloses are characterized by irregularity in cellulose chains composed from low molecular chains containing five member rings, open rings, and acidic parts. They are strongly hydrophilic and act as a component of cementing matrix between the cellulose microfibrils, forming the cellulose/hemicellulose network, which is considered to be the main structural component of the fiber cell. The lignin is hydrophobic on the other hand, acts as a cementing agent and increases the stiffness of the cellulose/hemicellulose composite.

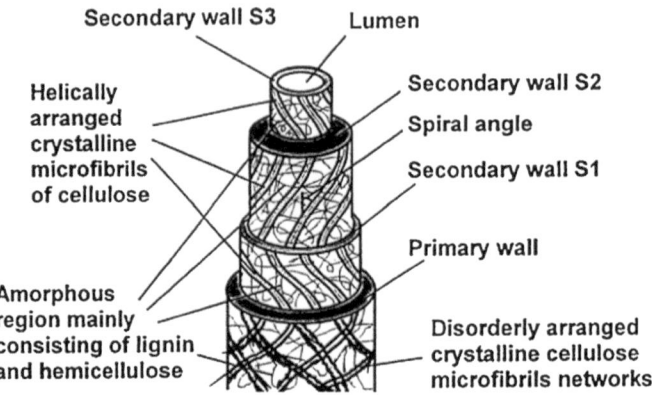

Fig. 2.4 Structural constitution of a natural cellulose fiber cell (Rong et al. 2001)

The plant fiber cell walls are divided into two main sections: a primary cell wall and a secondary cell wall. The primary cell wall consists of a loose irregular network of closely packed cellulose microfibrils, whereas the secondary wall is made up of three separate and distinct layers—S1 (outer layer), S2 (middle layer), and S3 (inner layer). S2 is the most important and thickest layer which determines the mechanical properties of fiber (Tsoumis 1991). Schematic illustration of the fine structure of a lignocellulosic fiber is presented in Fig. 2.4.

These fiber cell walls not only differ in the composition of cellulose, pectines, hemicellulose, and lignin but also in the orientation or microfibrillar/spiral angle of the cellulose microfibrils (Lee et al. 2014). Chemical composition, moisture content, and microfibrillar angle of some plant fibers are given in Table 2.1. The microfibrillar angle is the angle that the helical spirals of cellulose microfibrils form with the fiber axis. The microfibrillar angle varies from one plant fiber to another. The cellulose content in the fiber, microfibrillar angle, and the mean degree of polymerization of cellulose molecules are responsible for the mechanical properties of the fiber. Mean

Table 2.1 Chemical composition, moisture content, and microfibrillar angle of cellulose fibers (Mohanty et al. 2000c)

Fiber	Cellulose [wt%]	Hemicelluloses [wt%]	Lignin [wt%]	Pectin [wt%]	Moisture Content [wt%]	Waxes [wt%]	Microfibrillar Angle [°]
Flax	71	18.6–20.6	2.2	2.3	8–12	1.7	5–10
Hemp	70–74	17.9–22.4	3.7–5.7	0.9	6.2–12	0.8	2–6.2
Jute	61–71.5	13.6–20.4	12–13	0.2	12.5–13.7	0.5	8
Kenaf	45–57	21.5	8–13	3–5	–	–	–
Ramie	68.6–76.2	13.1–16.7	0.6–0.7	1.9	7.5–17	0.3	7.5
Sisal	66–78	10–14	10–14	10	10–22	2	10–22
Banana	63–64	10	5	–	10–12	–	–
Cotton	85–95	5.7	–	0–1	7.85–8.5	0.6	–
Coir	32–43	0.15–0.25	40–45	3–4	8		30–49

degree of polymerization also depends on the part of the plant from where the fibers are extracted. Fibers having higher cellulose content, higher mean degree of polymerization, and a lower microfibrillar angle display higher tensile strength and modulus.

Cellulosic fibers have both crystalline and amorphous domains. The crystallinity degree depends on the type and origin of the material. Cotton, flax, ramie, sisal, and jute have high degrees of crystallinity (65–70%), but the crystallinity of regenerated cellulose is only 35–40%. Progressive elimination of the less organized parts, i.e., amorphous domains, leads to fibrils with increasing crystallinity which can be almost 100% for cellulose whiskers. Crystallinity of cellulose results from the ordered arrangement of cellulose chains and from hydrogen bonding between them, but some hydrogen bonding also exists in the amorphous phase, although its organization is low (Lee et al. 2014). There are many hydroxyl (–OH) groups available in cellulose chains for interaction with water by hydrogen bonding. They interact with water at the surface as well as in the bulk. The quantity of water absorbed by the fiber depends on the relative humidity of the atmosphere. The sorption isotherm of cellulosic material depends on the degree of crystallinity and the purity of cellulose. All –OH groups in the amorphous region are easily accessible to water, whereas only a small amount of water interacts with the surface –OH groups of the crystalline region. The main components of plant-based natural fibers are cellulose (α-cellulose), hemicellulose, lignin, pectins, and waxes.

2.3.1 Cellulose

Cellulose is the major constituent of all plant fibers. Cellulose exists in polymer form of its β D-glucopyranose monomers that make a strong, rigid chain structure through polymerization of 1, 4-β glycosidic linkages. The monomers are linked in a state that one is turn over than other in each repeating unit. This gives the ability to the cellulose structure to make strong intramolecular and intermolecular hydrogen bonding due to the presence of hydroxyl groups as shown in Fig. 2.5. This produces a very compact and coherent structure that is responsible for the highly crystalline cellulose microfibrils. The strength of hydrogen bonds is very less as compared to the strength of covalent bonds but their presence in enormous amount in cellulose structure accounts for cellulose high structural strength. Although the chemical structure of cellulose for different plant fibers is same, the degree of polymerization and orientation of cellulose microfibrils varies considerably. The mechanical properties of a fiber are significantly dependent on the degree of polymerization and orientation of cellulose microfibrils.

2.3.2 Hemicelluloses

Hemicellulose like cellulose is a chain molecular substance but is distinguishable from the latter in having irregularities, branched chains containing pendant side groups, and

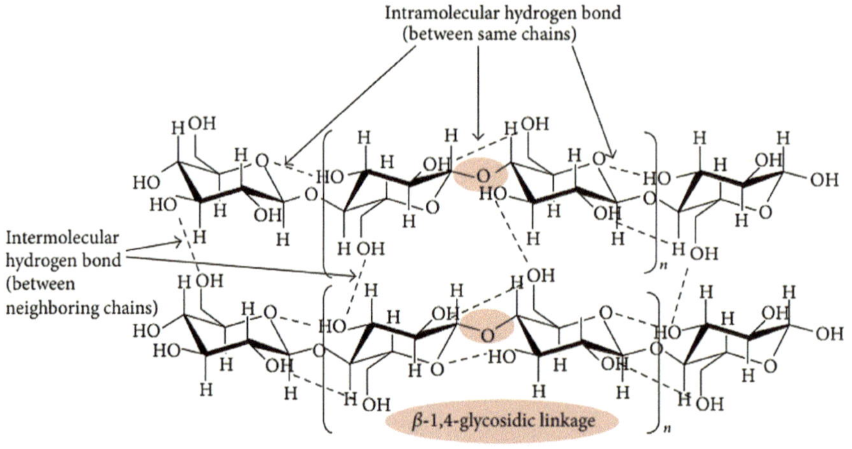

Fig. 2.5 Chemical structure of cellulose chains (Lee et al. 2014)

a relativity short chain length (low degree of polymerization) giving rise to amorphous nature. Hemicelluloses form the part of supportive matrix for cellulose microfibrils and are believed to be a compatibilizer between cellulose and lignin. Hemicellulose is very hydrophilic and soluble in alkali and easily hydrolyzed in acids. Hemicellulose occurs mainly in the primary cell wall and consists of polysaccharides of comparatively low molecular weight and built up from hexoses, pentoses, and uronic acid residues. It is mainly responsible for the biodegradation, moisture absorption, and thermal degradation of the fiber (Mohanty et al. 2000a). Figure 2.6 depicts the cell wall polymers, responsible for the properties of plant fibers in a better way.

2.3.3 Lignin

Lignin is a complex polymer which functions as the structural material and gives rigidity to the plant fibers. It is thought to be a complex, three-dimensional copolymer of aliphatic and aromatic constituents with very high molecular weight. Its chemistry has not yet been precisely established, but most of its functional groups and building units of the macromolecule have been identified. It is characterized by high carbon, but low hydrogen content. Hydroxyl, methoxyl, and carbonyl groups have been identified. Lignin has been found to contain five hydroxyl and five methoxyl groups per building unit. It is believed that the structural units of a lignin molecule are derivatives of 3-(4-hydroxy phenyl) prop-2-eneol. Lignin is amorphous and hydrophobic in nature. It is a thermoplastic polymer having a very slow thermal degradation which extends over the temperature range, starting from melting point 170 °C (Jonoobi et al. 2009). It is not hydrolyzed by acids, but soluble in hot alkali, readily oxidized and easily condensable with phenol. Lignin is thermally stable but is highly susceptible to

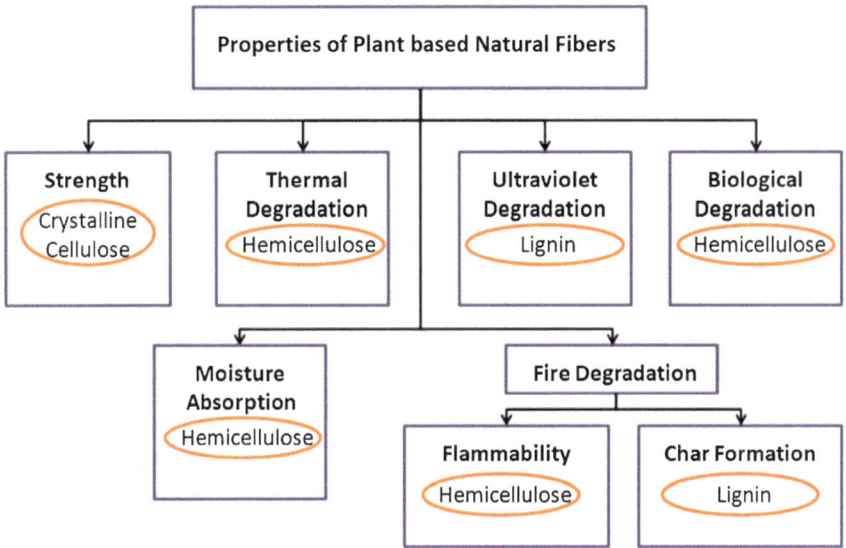

Fig. 2.6 Cell wall polymers responsible for the properties of plant fibers (Azwa et al. 2013)

ultraviolet light. Therefore, lignin is responsible for the ultraviolet light degradation of the fiber (Sedan et al. 2008) (Fig. 2.6).

2.3.4 Pectin

Pectin is a linear polysaccharide and mainly consists of D-galacturonic acid and corresponding methylester units joined in chains by means of 1, 4-α glycosidic linkage. The composition and structure of pectin are still not completely understood. Its structure is very difficult to determine because pectin can change during isolation from plant fibers, storage, and processing (Novosel'skaya et al. 2000). Additionally, impurities can accompany the main components.

2.4 Bast Fibers

Bast fibers constitute a significant share of the huge family of plant-based natural fibers. They are extracted from phloem/inner bark surrounding the stem of dicotyledonous plants. Figure 2.7 depicts the cross section of fibrous plant stem. Epidermis or skin protects the plant against moisture evaporation, sudden temperature changes and partly gives mechanical reinforcement to the stem of plant.

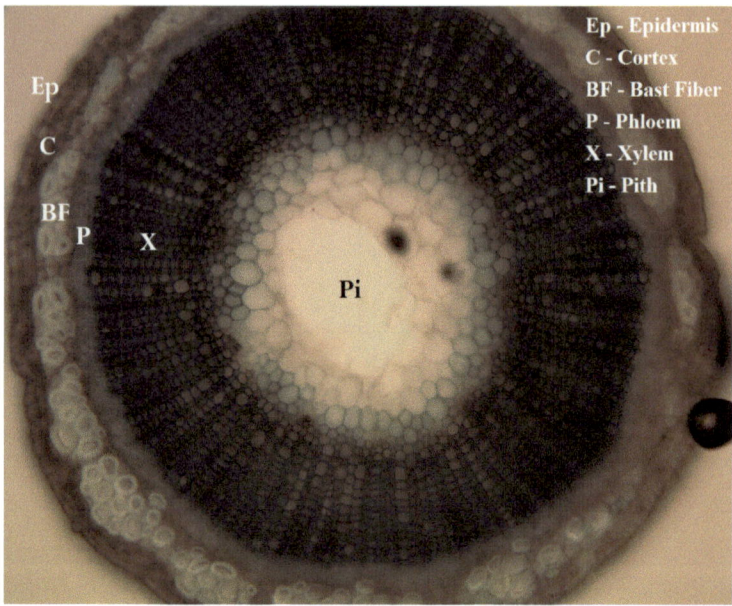

Fig. 2.7 Cross section depicting the position of bast fibers and other constituents of the bast fiber plant

Fibers are located in the phloem in the form of bundles under the skin. Xylem is the woody core in the middle part of the plant.

Bast fibers are separated from the woody core usually by water or dew retting and extracted by decortication. Retting is necessary to loosen the gummy substances which cement the fibers to the rest of the tissues in the stem and to each other. Decortication is the process of extraction of fiber bundles from retted stalk and is usually done manually. The fiber plants, longitudinal view and cross-sectional shapes of some important bast fibers are shown in Table 2.2. All bast fibers have lumen in their structure with different shapes and located in the central part parallel to fiber axis.

As bast fibers fall in the sub-category of plant/vegetable-based natural fibers, therefore, cellulose, hemicellulose, and lignin are the main constituents with 50–90% share of cellulose depending on the type of fiber and the part of stem from where the fiber is extracted. The properties of bast fibers are dependent on the conditions of cultivation and retting, varieties of fibrous plants as well as the condition of measurement. Bast fibers especially flax, hemp, jute, and kenaf have very good mechanical properties which are strongly related to the structure and composition.

Table 2.2 Typical bast fibers (Zimniewska et al. 2011)

Fiber	Botanical name	Plant photo	Longitudinal view of fibers	Image of fiber cross section
Flax	*Linum usitatissimum*			
Hemp	*Cannabis sativa*			
Kenaf	*Hibiscus cannabinus*			
Jute	*Corchorus capsularis*			
Ramie	*Boehmeria nivea*			

The structure, cell dimensions, microfibrillar angle, defects, and the chemical composition of fibers are the most important parameters that define the overall properties of the fibers (Satyanarayana et al. 1986). Generally, tensile strength and Young's modulus increase with higher cellulose content of fibers, higher degree of polymerization of cellulose, longer cell length, and lower microfibrillar angle. The microfibrillar angle is related to the stiffness of the fibers. The fibers are more ductile if the microfibrils have more spiral orientation to the fiber axis. If the microfibrillar angle is less and microfibrils are oriented parallel to the fiber axis, the fibers will be stiff, rigid, inflexible and have high tensile strength. The important mechanical properties of bast fiber are presented in Table 2.3 (Kabir et al. 2012).

Table 2.3 Comparative mechanical properties of bast and E-glass fibers (Kabir et al. 2012)

Fiber	Density [g/cm^3]	Tensile strength [MPa]	Young's modulus [GPa]	Specific strength [*GPa/g cm^{-3}]	Specific modulus [*GPa/g cm^{-3}]	Elongation at break [%]
Jute	1.3–1.4	393–773	13–26.5	0.3–0.5	10–18.3	1.16–1.5
Flax	1.50	345–1100	27–6	0.2–0.7	18.4	2.7–3.2
Hemp	1.48	690	30–60	0.6	26.3–52.6	1.6
Ramie	1.50	400–938	61.4–128	0.3–0.6	40.9–85.3	1.2–3.8
E-glass	2.5	2000–3500	70	0.8–1.4	28	2.5

*Stress divided by fiber density

2.5 A Brief Overview of Jute Fiber

Jute, also known as golden fiber, is the cheapest bast fiber. Jute is the second only to cotton in world's production of natural fibers. India, Bangladesh, Nepal, China, and Thailand are the leading producers of jute. It is also produced in southwest Asia and Brazil. More than 98% of total world production of jute is grown in three South Asian countries, i.e., Bangladesh, India, and Nepal. It belongs to the genus *Corchorus* and family *Tiliaceae*. There are over thirty *Corchorus* species but only two of them are widely known, *Corchorus capsularis* (white jute) and *Corchorus olitorius* (tossa jute). Jute is an important crop in Bangladesh and India and has good socioeconomic importance in these countries (Ranganathan and Quayyum 1993).

Jute can be cultivated under quite a wide variety of conditions but for ideal growth it requires a high level of humidity (40–97%). The ideal temperature lies between 17 and 41 °C. The pure fiber content of the unretted plants lies between 4.5 and 7.5%. Generally, after about 90–120 days of sowing, the stems may be harvested and water retted. Retting takes around 10–20 days and jute fibers are decorticated subsequently in the form of fiber bundles and washed and dried (Pan et al. 2000).

The longitudinal and cross-sectional view of jute fiber is shown in Fig. 2.8. The cross section shows polygonal shape with the canal (lumen) of different size comprising about 10% of the cell area of cross section. The fibers are coarse, generally 20–25 μm in diameter; the length of the ultimate fibers is only 2–5 mm. The cellulose in jute fiber has an average molecular weight between 130,000 and 190,000 with an average degree of polymerization of approximately 800–1200. Jute is a fairly strong fiber exhibiting brittle fracture but small extension at break and poor elastic recovery. The mechanical properties recorded in the literature vary considerably, may be due to variation in the linear density of fibers and differences in the methods of measurement.

(a)   (b)

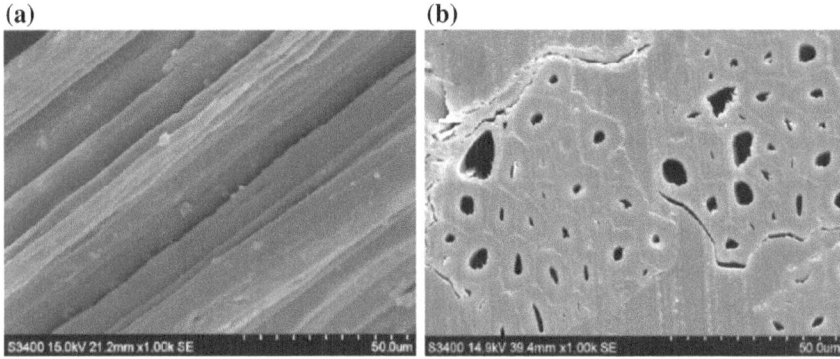

Fig. 2.8 a Longitudinal and **b** cross-sectional view of jute fiber (white jute) (Kicińska-Jakubowska et al. 2012)

2.6   Natural Fibers Reinforced Composites in Automotives

The concept of producing natural composite is about 3000 years old when clay reinforced with straw was used to build the walls of dwellings in ancient Egypt. However, natural materials emerged as a future possible material in early 1900s for use in the automotive sector (Suddell and Evans 2003). In 1941, during the World War II, natural fiber reinforced composites received considerable attention for making seats, bearings, and fuselages in aircraft due to shortage of aluminium at that time. The first example was "Gordon-Aerolite" a composite laminate of uni-directional flax yarn soaked with phenolic resin and cured under pressure, used as fuselages in aircraft. The other example was the cotton reinforced polymer composite, used by the military for aircraft radar (Piggott 1980). The first prototype composite car was developed from hemp fibers by Henry Ford in 1942 but unfortunately, the economic limitations hinder the general production of this car. Daimler-Benz has been working on the idea of replacing glass fibers with natural fibers in automotive components since 1991. In 1996, jute-based door panels were introduced by Mercedes-Benz into its E-class vehicles. The door trim panels developed from hybrid flax/sisal mate reinforced polyurethane composites were used in Audi A2 midrange car in 2000 (Mohanty et al. 2005). All body panels of a small prototype car were manufactured and assembled by a researcher in Brazil using jute fiber reinforced composites and hybrid composites (Al-Qureshi 2001) as depicted in Fig. 2.9. A remarkable 20% weight reduction is made in E-class Mercedes-Benz car using interior components made from a blend of flax and sisal fibers in an epoxy matrix (Jawaid and Khalil 2011) as shown in Fig. 2.10. Moreover, natural fiber reinforced composites with a total weight of 43 kg were used to manufacture 27 components in Mercedes S-class car (Pickering 2008).

Almost all well-known car manufacturers in the Europe are now using natural fiber composites in various interior components as those listed in Table 2.4. In reference to European Union (EU) guideline 2000/53/EG on the end of life vehicle

Fig. 2.9 A prototype car made from jute fiber reinforced composite and hybrid composite in Brazil (Al-Qureshi 2001)

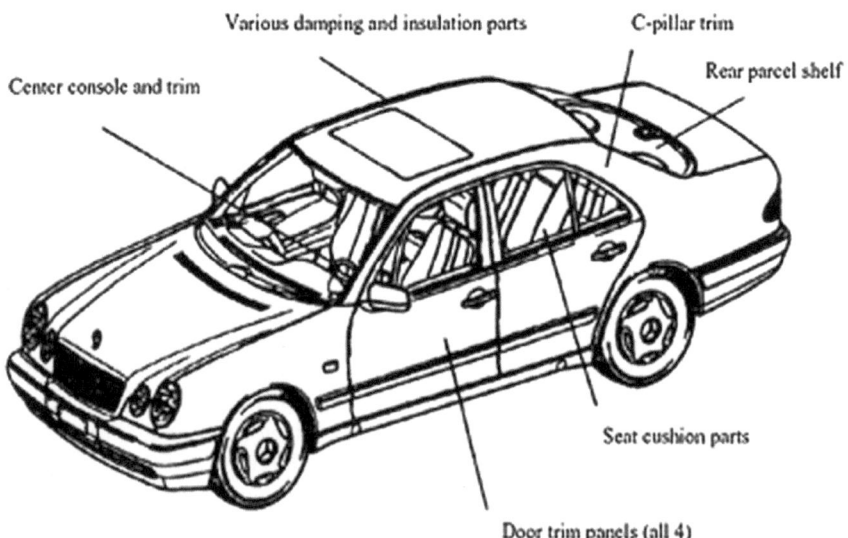

Fig. 2.10 Natural fiber composites application in the current E-Class Mercedes-Benz (Jawaid and Khalil 2011)

Table 2.4 Vehicle Manufacturers and use of natural fiber composites

Automotive manufacturer	Model	Applications
Audi	A2, A3, A4 (and Avant), A6, A8	Roadster, coupe, seat backs, side and back door panels, boot lining, hat racks, spare tyre lining
BMW	3, 5, 7 series	Door panels, headliner panel, boot lining, seat backs, noise insulation panels, molded foot well linings
Citroën	C5	Interior door panels
Daimler-Chrysler	A, C, E, and S-class;	Door panels, wind shield, dashboard, business table, pillar cover panel
Ford	Mondeo CD 162, Focus	Door panels, B-pillar, boot liner
Lotus	Eco Elise	Body panels, spoiler, seats, interior carpets
Mercedes-Benz	Trucks	Internal engine cover, engine insulation, sun visor, interior insulation, bumper, wheel box, roof cover
Peugeot	406	Seat backs, parcel shelf
Renault	Clio, Twingo	Rear parcel shelf
Vauxhall	Corsa, Astra, Vectra, Zafira	Headliner panel, interior door panels, pillar cover panel, instrument panel
Volkswagen	Golf, Passat, Bora	Door panel, seat back, boot lid finish panel, boot liner
Volvo	C70, V70	Seat padding, natural foams, cargo floor tray
Rover	2000 and others	Insulation, rear storage shelf/panel

(ELV) issued by the European Commission, 95% of the weight of a vehicle have to be recyclable by 2015 with 85% recoverable through reuse or mechanical recycling (Peijs 2003).

The car manufacturers in Germany are striving to make every component of their vehicle either recyclable or biodegradable (Hill 1997). In order to produce fuel efficient and low polluting vehicles, natural fiber composites are considered the ideal replacement of glass fiber reinforced plastics (GFRP) where appropriate, because of the main advantages of reduction in cost and weight. Currently, natural fibers account to over 14% share of reinforcement materials; however, the share is projected to rise to 28% by 2020 (HOBSON and CARUS 2011).

2.7   Polymer Matrices

Polymer matrix in a composite provides uniform load distribution to the reinforcing fibers and holds them together in place. They are usually of lower strength compared to the reinforcing fibers. Additionally, the matrix safeguards the composite surface against abrasion, mechanical damage, and environmental corrosion (Akovali 2001). The polymer matrix should be strong enough to withstand the load but also good enough to transfer load to the reinforcing fibers. The major categories of polymer matrices are thermosets, thermoplastics, rubber matrices, and biobased polymer matrices.

2.7.1 Thermosets

Thermoset matrices are the most frequently used matrix materials in polymer-based composites industry, mainly because of their ease of processing. They are low molecular weight reactive oligomers at the beginning. Generally, they contain two (telechelic oligomer and curing agent) or more components and solidification begins when the components are mixed either at ambient or elevated cure temperatures. The subsequent reaction produces a rigid, highly crosslinked network or a vitrified system with exceptional strength (Saheb and Jog 1999). Epoxies, vinyl esters, polyesters, phenolic resins, and polyurethanes account for the majority of thermoset resins used in industry.

2.7.2 Thermoplastics

Thermoplastics are heat softenable, heat meltable, and reprocessable having one- or two-dimensional molecular structures as opposed to three-dimensional structures of thermosets. They usually come in the form of molding compounds that soften at high temperatures and consist of linear or branched chain molecules having strong intramolecular bonds, but weak intermolecular bonds (Pickering 2008). Their structure is either semicrystalline or amorphous. Melting and solidification of these polymers are reversible, and they can be reshaped by application of heat and pressure. Thermoplastic materials that currently dominate as matrices are polypropylene (PP), polyethylene (PE), polystyrene (PS), polyether-ether ketone (PEEK), and poly (vinyl chloride) (PVC). One of the limitations is the need to process the thermoplastic composites below the decomposition temperature of cellulose, which is $\sim 190\ °C$. Only PP and PE matrices are amenable to natural fiber reinforcement. Of all the thermoplastics, PP shows the most potential benefits when combined with natural fibers for making biocomposites of industrial value. Among all the thermoplastics, the PP matrix natural fiber biocomposites show the most potential benefits of industrial value.

2.7.3 Rubber Matrices

The main classes of rubber matrices that have been used for the preparation of composites are: natural rubber (NR), butyl rubber (IIR), butadiene rubber (BR),

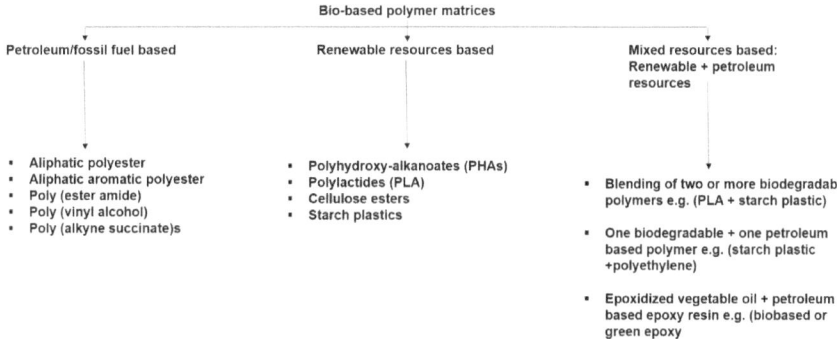

Fig. 2.11 Broad classification of biobased polymer matrices

styrene butadiene rubber (SBR), nitrile rubber (NBR), chloroprene rubber (CR), ethylene propylene diene rubber (EPDM), polyurethane rubber, and silicon rubbers but the most widely used rubber matrix is natural rubber.

2.7.4 Bio-based Polymer Matrices

The US Department of Agriculture and the US Department of Energy have set goals of having at least 10% of all basic chemical building blocks be created from renewable, plant-based sources in 2020, increasing to 50% by 2050 (Dittenber and GangaRao 2012). Currently, numerous researches are underway to develop a new class of composites, known as "green composites" by combining natural fibers with biodegradable/biobased resins. The classification of biobased polymers is presented in Fig. 2.11. Biobased polymers may or may not be fully biodegradable, depending on their structure, composition, and on the environment in which they are placed (Mohanty et al. 2005). Therefore, there is an ambiguity in the definition of biodegradable or green and biobased polymers. Most of the biodegradable epoxy polymers are not completely biobased nevertheless, there has been the development of oxidized green polymers from natural oils. As biobased green epoxy is used in the current study; it is the petroleum-derived epoxy resin blended with epoxidized vegetable oil in the presence of suitable curing agent.

2.8 Jute Fiber Reinforced Polymer Composites

Jute is considered the potential bast fiber for reinforcement in composites due to its good mechanical properties, cheaper availability, biodegradability, and large production relative to other bast fibers. Ray et al. (2002a, b) and (2004) (Sarkar and Ray 2004) extensively investigated the effect of alkali treatment on the mechanical, dynamical mechanical, thermal, and impact fatigue properties of jute/vinyl ester composites. The results revealed that longer alkali treatment was more helpful to remove hemicelluloses and to improve the crystallinity of fibers thus enabling better fiber dispersion. The mechanical, dynamic mechanical, thermal, and impact properties were superior owing to the alkali treatment, which comprises treatment time, concentration, and conditions. In another study (Sudha and Thilagavathi 2016), the effect of alkali treatment on the jute fabrics and its influence on various mechanical properties such as tensile, flexural, and impact strength of jute/vinyl ester composites was studied. Alkali-treated samples exhibited the improvement in mechanical properties of composites which may be due to better adhesion between the fabric and the resin because of the removal of lignin and hemicellulose. Mechanical properties of alkali-treated nonwoven jute felt reinforced soy composites exhibited better tensile properties than those of raw jute felt composites (Avancha et al. 2013). Gassan and Bledzki (1999) also investigated the influence of alkali treatment on the mechanical properties of jute fibers as well as jute/epoxy composites. The strength and stiffness of composites were increased as a result of the improved mechanical properties of the fibers after alkali treatment.

The effect of surface modification of jute fabrics on the mechanical and biodegradability of jute/Biopol composites was studied (Mohanty et al. 2000b). The tensile strength was found to improve by more than 50%, bending strength by 30%, and impact strength by 90% in the composites as compared to values achieved for pure Biopol sheets. Degradation studies showed that more than 50% weight loss of the jute/Biopol composites occurs after 150 days of composite burial.

Gao and Mäder (2006) studied jute fiber reinforced polypropylene (PP) composites to evaluate the effect of matrix modification using maleic anhydride (MAH) graft copolymer and revealed the significant improvement in the adhesion strength with jute fibers and in turn the mechanical properties of composites. Gassan and Bledzki (1997) also studied the effectiveness of MAH graft copolymer on the mechanical properties of jute-PP composites. Flexural and dynamic strength of MAH-PP treated composites were increased due to improvement of fiber/matrix adhesion. Jute fiber reinforced polypropylene composites were evaluated regarding the influence of gamma radiation (Khan et al. 2009). Mechanical properties such as tensile strength, tensile modulus, bending strength, bending modulus, and impact strength of the gamma irradiated composites were found to be higher than that of untreated composites. The effect of interfacial adhesion on creep and dynamic mechanical behavior (Acha et al. 2007), the influence of silane coupling agent (Hong et al. 2008; Wang et al. 2010), the effect of natural rubber (Zaman et al. 2010) on the mechanical properties and the effect of potassium permanganate on the mechanical, thermal, and degradation properties (Khan et al. 2013) of jute-PP composites were also explored by different researchers.

Different thermoset plastic resins were used as matrices for jute fiber reinforced composites and properties including the thermal stability (Sarkar and Adhikari 2001), mechanical and thermomechanical behavior (Abdallah et al. 2010), durability (Singh et al. 2000), fiber orientation on frictional and wear behavior (Dwivedi and Chand 2009), eco-design of automotive components (Alves et al. 2010), and alkylation effect on tensile, flexural and interlaminar shear strength (ILSS) (Sarikanat 2009) were examined.

The properties of jute fiber reinforced polyester composites were studied, including the relationship between water absorption and dielectric behavior (Fraga et al. 2006), impact damage characterization (Santulli 2001), weathering and thermal behavior (Dash et al. 2000a), effect of silane treatment on mechanical properties (Sever et al. 2010), effect of enzyme treatment on dynamic mechanical and thermal behavior (Karaduman and Onal 2013), and influence of copper incorporation on the mechanical and thermal behavior (Biswas et al. 2016).

The mechanical properties of PLA were improved significantly with jute reinforcement (Plackett et al. 2003). A 40 wt% composite of jute-PLA had doubled the strength of a pure PLA sample, though the impact resistance between the samples did not differ. The increase in tensile strength was temperature dependent and was contingent on the heating stage during composite formation not exceeding 210–220 °C. Tao et al. (2009) found that the mechanical properties of jute-PLA composites were optimum with 30% fiber volume fraction, and thermogravimetric analysis of composites showed that the addition of fiber to the composite improved the degradation temperature. Hongwei et al. (Ma and Joo 2011) reported that the optimum tensile properties of jute-PLA composites were obtained at 15 wt% fiber content and a processing temperature of 210 °C, whereas the maximum flexural strength and modulus of composites were obtained at 220 °C and 15 wt% fiber contents. The effect of different fiber surface treatments such as alkali, permanganate, peroxide, and silane on mechanical and abrasive wear performance of jute-PLA composites was studied (Goriparthi et al. 2012). Surface treatments resulted in enhancement of tensile and flexural properties and reduction in Izod impact strength of composites. The dynamic mechanical behavior exhibited higher storage modulus and lower tangent delta of treated composites than that of untreated composite and silane-treated composite showed higher thermal stability.

2.9   Micro-/Nanocellulose Filler Composites

Cellulose is an abundant low cost and renewable natural polymer exists in nature and is a major structural component of plant cells. The two types of nanofiller for composites, obtained from cellulose, are cellulose microfibrils and cellulose nanocrystals (CNC) or cellulose nanowhiskers (CNW) (Azizi Samir et al. 2005). Both chemical and mechanical means are usually adopted to extract cellulose fibrils and nanowhiskers from cellulosic resources. The wood of forest resources, lignocellulosic fibers (hemp, flax, jute, ramie, kenaf, etc.), and agricultural residues or by-products (corncob, risk husk, sugarcane bagasse, etc.) are the abundant, cheaper, and readily available cellulose plant resources around the world (Ng et al. 2015). The extracted cellulose fillers have a perfect crystalline structure (about 65–95% crystallinities) with considerable characteristics such as their abundant hydroxyl groups, high aspect ratio, large specific surface area, good mechanical properties, and high thermal stability which make them a good choice as polymer reinforcing filler in composites (Ng et al. 2015). It has been observed that cellulose crystals extracted from cellulose resources have tensile strength around 10 GPa and modulus around 150 GPa (Helbert et al. 1996) which suggest that cellulose can replace single-walled carbon nanotubes (SWCNTs) in many applications.

There are numerous methods that can be used to produce large quantities of cellulose fibrils/crystals in the laboratories as reviewed in the study by Siró and Plackett (2010). The aspect ratio, orientation, distribution, and loading of cellulose fillers in the polymer matrix decide the properties of composites (Jiang et al. 2007; Kvien and Oksman 2007). A lot of work can be found in the literature on nanocellulose-filled polymer composites, but only a few of them are discussed here. Xu et al. (2013) reported the mechanical and thermomechanical properties of waterborne epoxy reinforced with cellulose nanocrystals. The tensile strength, tensile modulus, glass transition temperature (T_g), and storage modulus increased with increasing filler content, indicating good reinforcement of the epoxy resin matrix. Isora nanofibrils (INFs) reinforced unsaturated polyester nanocomposites were studied for their mechanical and viscoelastic behavior (Chirayil et al. 2014). The improved network in the polyester matrix due to higher aspect ratio of isora fibrils resulted in the improvement of mechanical, T_g, and water barrier properties of prepared composites. Cellulose nanowhisker prepared by acid hydrolysis of microcrystalline cellulose (MCC) were used as filler to reinforce poly(vinyl alcohol) (PVA) with 1, 3, 5, 7 wt% loadings in order to evaluate the mechanical and thermal properties of nanocomposites (Cho and Park 2011). The tensile strength, modulus, and thermal stability were found to increase with the increase in nanocellulose content. However, DMA result showed a significant increase of the storage modulus of the nanocomposite at the 3 wt% of nanocellulose. Jonoobi et al. (2010) used kenaf pulp to isolate cellulose nanofibers which were reinforced with polylactic acid at 1, 3, 5 wt% loadings of filler. The results showed the improvement in tensile strength and modulus of composites with the increase in loading of cellulose fillers. The storage modulus was also increased for all samples as compared to neat PLA and tan delta peak shifted to higher temperature. The work on hybrid carbon woven fabric/epoxy composites reinforced with two types fillers, namely microfibrillated cellulose (MFC) and carboxylated nitrile-butadiene rubber nanoparticles (XNBR), has been reported recently to evaluate their fatigue performance (Carvelli et al. 2016). The best fatigue life was shown by the composites with maximum content of MFC in the system.

2.10   Surface Treatments of Natural Fibers

The fiber/matrix interphase plays an important role to characterize the mechanical properties of the composites. Strong interfacial adhesion is responsible for the good mechanical properties of composites. Several problems occur at the interphase when natural fibers are used as reinforcement in polymer composites due to the presence of hydrophilic hydroxyl groups at the surface of natural fibers as this hydrophilic nature hinders the effective reaction with the matrix. Additionally, the presence of pectin and waxy substances cover the reactive functional groups of the fiber and act as a barrier to interlock with the matrix. Therefore, different chemical, physical, and biological methods (which are discussed briefly in the following sections) are used for surface modification of natural fibers in order to enhance the fiber/matrix interfacial interaction.

2.10.1   Chemical Treatment of Natural Fibers

Chemical methods of fiber treatment help to expose more reactive groups on the fiber surface thus facilitating efficient coupling with the matrix and hence better mechanical properties of composites (Dash et al. 2000b). These methods actually reduce the hydrophilic tendency of natural fibers and thus improve the compatibility with the matrix. The following are some of the most commonly used methods for chemical modification of natural fibers.

2.10.1.1   Alkali Treatment

Alkali treatment of natural fibers is a widely used method to modify the cellulosic molecular structure. It changes the orientation of highly packed crystalline cellulose. In fact, alkali has a swelling reaction on a cellulosic fiber, during which the natural crystalline structure of the cellulose relaxes. The native cellulose in natural fibers has a monoclinic crystalline lattice of cellulose-I which can be changed into different polymorphous forms through chemical or thermal treatments. The important transformations are alkali-cellulose and cellulose-II as shown in Fig. 2.12. The type of alkali (KOH, LiOH, NaOH) and its concentration influence the degree of swelling and hence the degree of lattice transformation into cellulose-II (Fengel and Wegener 1983). Sodium hydroxide is more effective in cellulose swelling due to favorable diameter of Na^+ which is able to penetrate the smallest pores in between the lattice planes. Cellulose micromolecules are separated at large distances due to swelling and alkali sensitive hydroxyl (OH) groups present among the molecules are broken down, react with water molecules and then move out from the fiber structure.

The remaining reactive molecules form fiber–cell–O–Na groups between the cellulose molecular chains (John and Anandjiwala 2008). Hence, the hydrophilic hydroxyl groups are reduced and the fibers moisture resistance property is increased. Afterward, rinsing with water removes the linked Na-ions and converts the native cellulose to a new crystalline structure, i.e., cellulose-II.

This treatment also removes a certain portion of hemicelluloses, lignin, pectin, wax, and oil covering materials. The surface roughness is increased and fiber

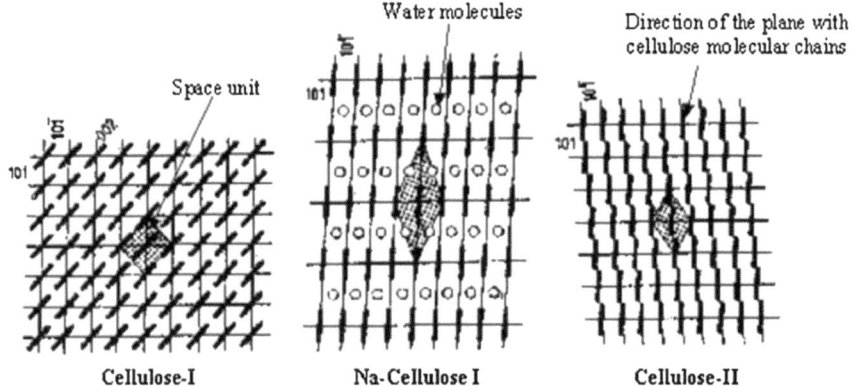

Fig. 2.12 A schematic representation of the transformations of crystalline lattices of cellulose-I, Na-cellulose-I, and cellulose-II by alkali treatment (Van de Weyenberg et al. 2006)

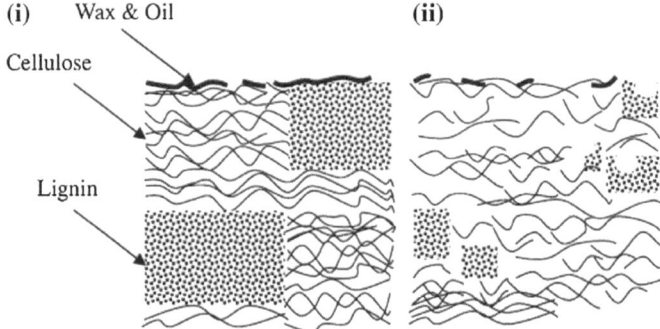

Fig. 2.13 Typical structure of (*i*) untreated and (*ii*) alkali-treated cellulose fiber (Kabir et al. 2012)

diameter is reduced, thereby aspect ratio (length/diameter) is increased. Hence, effective fiber surface area becomes larger for good adhesion with matrix. The alkali concentration higher than the optimum value can also damage the cellulose structure which can adversely affect the fiber and composite properties. A schematic view of natural fiber structure before and after alkali treatment is shown in Fig. 2.13 (Mwaikambo and Ansell 2002).

2.10.1.2 Silane Treatment

Silane is a multifunctional chemical compound and used as a coupling agent to modify the surface of fibers. Silane helps to form a chemical linkage between the fiber surface and the matrix through a siloxane bridge. Silanes may reduce the number of hydroxyl groups in the fiber structure. In the presence of moisture, silanols form due to hydrolysable alkoxy groups (Sreekala et al. 2000). One end of the silanol reacts with the hydroxyl groups of fiber cell and other end reacts with the matrix functional group thus providing molecular continuity across the interphase of the composite.

2.10.1.3 Acetylation Treatment

Acetylation treatment describes an esterification method in which an acetyl functional group (CH_3COO^-) is introduced into an organic compound. In case of natural fibers, this acetyl functional group passivates hydroxyl groups of fiber and takes hydrophobic nature. The hydrophilic nature of fiber results in the dimensional stability of composites. Usually it is not done independently, but is preceded by pretreatment with alkali.

2.10.1.4 Benzoylation Treatment

Benzoyl chloride is mostly used in benzoylation treatment of natural fibers in order to reduce their hydrophilic nature and to improve interfacial adhesion in composites. Alklai pretreated fibers are treated with benzoyl chloride. During the reaction, benzoyl group is attached to the cellulose backbone by replacing hydroxyl groups resulting in a more hydrophobic nature of fiber (Joseph et al. 2003).

2.10.1.5 Peroxide Treatment

The functional group of peroxide can be represented as ROOR. Most commonly used peroxides for the treatment of natural fibers are benzoyl peroxide and dicumyl peroxide. The main advantage of peroxide treatment is the quick decomposition of a peroxide yielding free radical that can react with the hydroxyl group of the fiber and with the matrix resulting in good fiber/matrix adhesion along the composite interphase. Like acetylation and benzoylation treatments, fibers are pretreated with alkali before treating with peroxides. Higher temperature is more suitable for achieving the complete decomposition of peroxide (Sreekala et al. 2000).

2.10.1.6 Permanganate Treatment

Potassium permanganate $(KMnO_4)$ in acetone solution is mostly used for permanganate treatment of natural fibers. The fibers are soaked in solution and concentration is carefully controlled to form highly reactive Mn^{2+} ions that react with the cellulose hydroxyl groups and form cellulose–manganate for initiating graft copolymerization. The hydrophilic tendency of fibers after permanganate treatment is reduced and chemical interlocking at the interphase is enhanced providing better fiber/matrix adhesion (Rahman et al. 2007).

2.10.1.7 Isocyanate Treatment

The isocyanate functional group (–N=C=O) of isocyanate compound reacts with the hydroxyl group of cellulose and lignin constituents of the fiber thus forming a urethane linkage which provides strong covalent bonds between fiber and the matrix. Additionally, isocyanate reacts with moisture present in the fiber and forms urea which reacts further with hydroxyl groups of fiber constituents, thus reducing the hydrophilic tendency of fiber (George et al. 2001).

2.10.1.8 Maleated Coupling Agents

Maleic anhydride coupling agent is frequently used to modify the fiber surfaces and provides efficient interaction with the functional surface of the fiber and matrix. This method is mostly used to modify fiber surfaces destined for pairing with a polypropylene matrix. Maleic anhydride reacts with the hydroxyl groups in the amorphous region of fiber structure. This reaction produces brush like long chain polymer coating on the fiber surface and reduces its hydrophilic nature (George et al. 2001). The covalent bonds between the hydroxyl groups of the fiber and the anhydride groups of maleic anhydride make bridge interface for efficient inter-locking (Keener et al. 2004).

2.10.1.9 Graft Copolymerization

Acrylic acid (CH_2=CHCOOH), acrylonitrile (CH_2=CH–C≡N), and vinyl mono-mers are mostly used for graft copolymerization of natural fibers. Free radicals, initiated on cellulose molecule during grafting, interact with monomer of matrix thus enhancing the interlocking efficiency at the interphase (Kalia et al. 2009).

2.10.2 *Physical Treatments of Natural Fibers*

Natural fibers can be surface modified by physical methods which include corona, plasma, steam explosion, and high energy irradiation (laser, UV and gamma rays etc.). The ultimate purpose of all these methods is to improve the fiber/matrix adhesion by changing the structural and surface properties of the fiber. All physical treatments are considered the eco-friendly ones. The chemical composition of the fibers is not extensively changed by physical treatments. Therefore, the mechanical bonding between the fiber and the matrix is mainly responsible to enhance the interphase properties of composites. A brief description of corona, plasma, and steam explosion treatments is given below.

2.10.2.1 Plasma Treatment

Plasma is defined as a gaseous environment composed of charged and neutral species with an overall zero charge density (Kalia et al. 2013). Plasma treatment changes the surface properties of the fibers through the formation of free radicals, ions, and electrons in the plasma stream which can result in surface cleaning, etching, roughness or activation depending on the type and nature of used gases like helium, oxygen, air, and nitrogen. Low temperature, low pressure, atmospheric pressure plasma, and atmospheric glow discharge treatments are mostly used to modify the surface of natural fibers (Kalia et al. 2013). A schematic representation of atmospheric air pressure plasma is shown in Fig. 2.14 in which a high frequency electric current excites a feeding gas usually compressed air, into relatively low temperature plasma (Baltazar-y-Jimenez et al. 2008).

2.10.2.2 Corona Treatment

Corona discharge treatment (air plasma treatment) is an effective method of surface modification to enhance the fiber/matrix interaction in composites. This process changes the surface energy of the cellulose fibers by surface oxidation activation (Faruk et al. 2012). The corona plasma is generated by the application of high voltage to an electrode. The principle of corona discharge system is shown in Fig. 2.15.

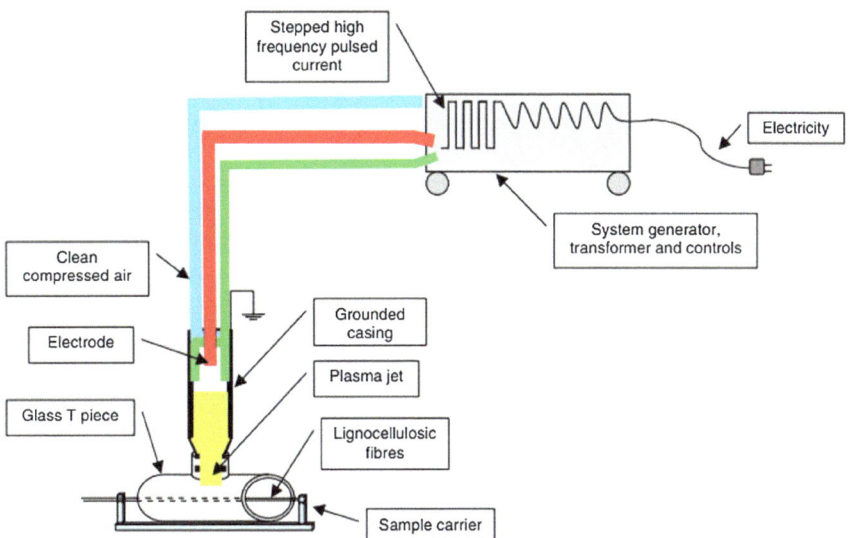

Fig. 2.14 A schematic view of atmospheric pressure plasma treatment (Baltazar-y-Jimenez et al. 2008)

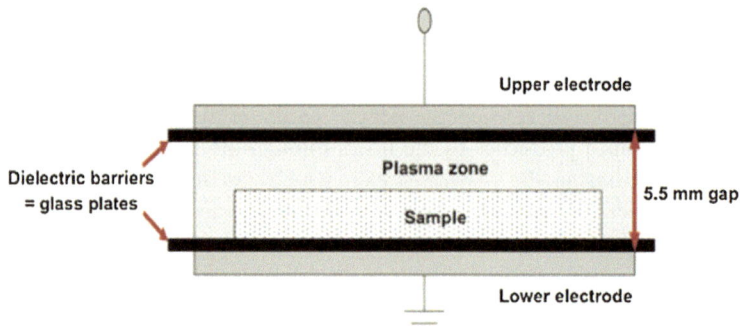

Fig. 2.15 Principle of corona discharge system (Ragoubi et al. 2010)

2.10.2.3 Steam Explosion

In steam explosion process, a high-pressure shock steaming of lignocellulosic materials at high temperature and pressure is involved followed by mechanical disruption of the pretreated material by violent discharge (explosion) into a collecting tank. This process has been applied to many lignocellulosic materials including plant fibers, to enhance dispersibility and adhesion with the polymer matrix (Satyanarayana 2004).

2.10.3   Biological Treatments of Natural Fibers

Biological agents such as enzymes, bacterial cellulose, and fungi provide an alternative way to chemical and physical methods for surface modification of natural fibers. These methods are also eco-friendly; therefore, the use of these methods is rapidly expanding. Biological methods offer several advantages over chemical and physical ones. They can selectively remove pectin and the hemicelluloses while requires less energy input (Gurunathan et al. 2015).

2.10.3.1 Enzyme Treatment

Enzymes offer an environment-friendly alternative to chemical methods for surface modification of natural fibers and their use is becoming increasingly substantial. Surface modification by the enzyme is safer and more advantageous as compared to chemical methods because of high reaction specificity of enzymes, milder reaction conditions, and non-destructive transformations on the surface of fiber (Pallesen 1996). The action of enzyme on the fiber cell is presented in Fig. 2.16 (Kalia et al. 2013). Surface modification by enzyme (usually by cellulase) results in the

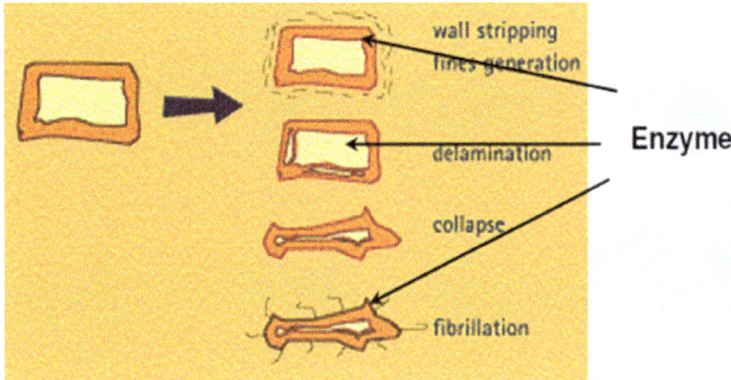

Fig. 2.16 Action of enzymes on the plant cell (Kalia et al. 2013)

degradation of cellulose in the fiber cell wall structure thus initiating the wall stripping, causing the generation of fine fibrils, and leaving the fibers less hydrophilic.

2.10.3.2   Fungi Treatment

Treatment of natural fibers with the fungi is also an eco-friendly and efficient alternative method to the chemical methods. Fungi increase the solubility of hemicelluloses thus reducing the hydrophilicity of the fiber. Fungal treatment also helps to remove lignin from natural fiber and causes the formation of holes (pits) on fiber surface, which provides roughness to the fiber surface and ultimately increases the interfacial adhesion between fiber and matrix (Jafari et al. 2007; Pickering et al. 2007; Kabir et al. 2012). Before treatment with fungi, the fibers are sterilized in an autoclave at 120 °C for 15 min. Then, fungi are added to the fibers with certain proportion and incubated at 27 °C for 2 weeks. Afterward, the fibers are sterilized, washed, and oven dried. The use of white rot fungus can be found in the literature for the treatment of plant fibers (Pickering et al. 2007).

2.10.3.3   Treatment with Bacterial Nanocellulose

The coating of natural fibers with bacterial cellulose involves the deposition of nanosized cellulosic materials onto the surface of fibers to enhance the interfacial adhesion between the fiber and the matrix (Lee et al. 2011). The bacterial cellulose is favorably deposited in situ onto the surface of natural fibers when cellulose producing bacteria such as *A. xylinum* is cultured in the presence of fibers in an appropriate culture medium. This deposition of bacterial cellulose onto fibers provides a new way of controlling the interaction between the fiber and the matrix.

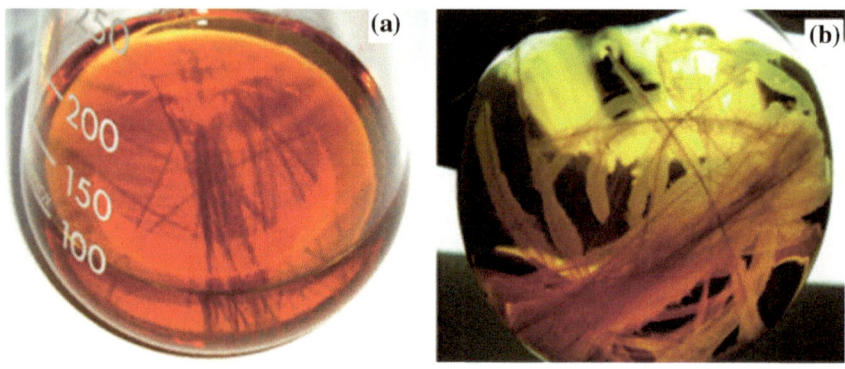

Fig. 2.17 Natural fibers **a** before and **b** after 2 days of bacterial treatment (Kalia et al. 2013)

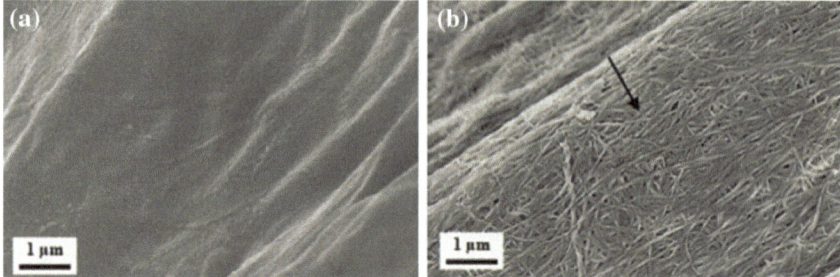

Fig. 2.18 Microscopic images showing **a** neat sisal fiber **b** sisal fiber coated with bacterial cellulose (Lee et al. 2011)

This method facilitates good distribution of bacterial cellulose within the matrix which results in an improved interfacial adhesion between fibers and the matrix through mechanical interlocking.

Figure 2.17 shows the culture medium and natural fibers immersed in the culture medium before and 2 days after culturing (Kalia et al. 2013). A layer of bacterial cellulose pellicles can be seen growing away from the surface of the natural fibers. The surface of sisal fibers was successfully modified by culturing cellulose producing bacteria in the presence of fibers in an appropriate culture medium as shown in Fig. 2.18 (Lee et al. 2011).

2.11 Creep in Natural Fiber Composites

Creep is defined as "a progressive deformation in a material under constant applied load". All polymeric materials including polymer composites undergo creep deformation even at room temperature. Creep is unwanted phenomena especially in

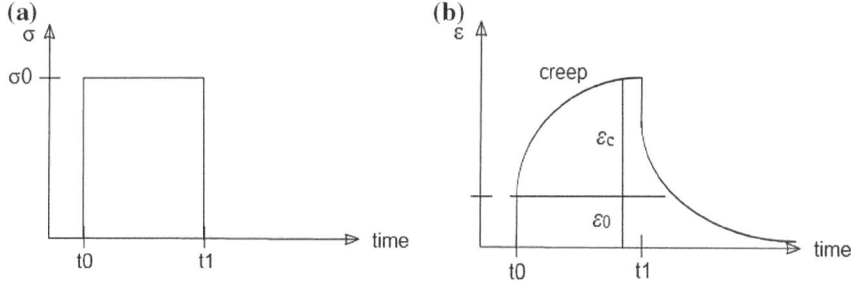

Fig. 2.19 Creep **a** application of constant stress; **b** strain response

composites because it is source of time dependent form of instability in loaded structures. The creep deformation is realized under a constant applied load as shown in Fig. 2.19. An instantaneous strain (ε_0), proportional to the applied stress, is observed after the application of constant stress (σ_0), followed by a continuous increase in strain as shown in Fig. 2.19b. The total strain (ε_t) at any instant of time t is represented as the sum of the instantaneous elastic strain (ε_0) and the creep strain (ε_c), i.e.,

$$\varepsilon(t) = \varepsilon_0 + \varepsilon_c. \tag{2.1}$$

Generally, creep can be described in three stages: primary, secondary, and tertiary. During the primary stage, the material undergoes deformation at a decreasing rate, followed by a region where deformation occurs at a nearly constant rate, whereas in the tertiary stage, it occurs at an increasing rate and ends with fracture as depicted in Fig. 2.20.

Creep phenomenon is a very complex and depends on many material parameters. For example, in case of fiber reinforced composites, creep behavior depends not only on the viscoelastic response of the matrix and the fiber but also on the elastic and fracture behavior of the fibers, geometry and arrangement of the fibers, and the fiber/matrix interfacial properties. Fiber reinforced polymer composites

Fig. 2.20 Stages of creep (Betten 2008)

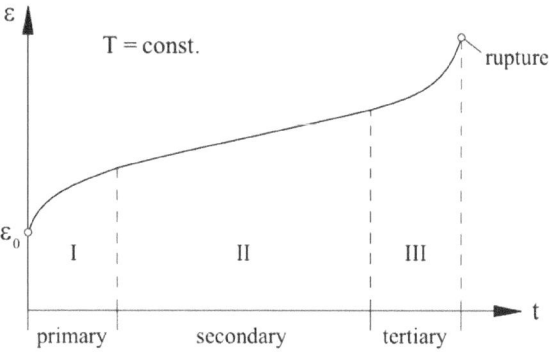

used in engineering applications are often subjected to stresses for a long time and at high temperatures in some situations as well. Creep is therefore a very important parameter in engineering design and manufacturing as this can lead to dimensional instability of the end product or even failure at applied constant stresses that are significantly lower than the ultimate tensile strength of material.

Considerable studies can be found in the literature on the creep behavior of natural fiber polymer composites and nanofiller composites. Nuñez et al. (2004) investigated the short-term and long-term creep behavior of woodflour/polypropylene (PP) composites at different temperatures using polypropylene maleic anhydride copolymer (PPMAN) as compatibilizer between the filler and matrix. The results were discussed with reference to effects of filler content, addition of compatibilizing agent and temperature. The study resulted in a reduction of creep deformation with the addition of woodflour and PPMAN in the matrix and decrease in temperature. Acha et al. (2007) also studied the effect of interfacial adhesion between jute fabric and polypropylene on the creep behavior of composites by chemically treating the jute and using two commercial maleated coupling agents. The time-temperature superposition principle (TTSP) was used to predict the long-term creep performance. The creep deformation was directly related to the interfacial properties of composites. The effect of temperature on the creep and recovery behavior of kenaf nonwoven fabric reinforced polypropylene composites was investigated in another study (Hao et al. 2014), and long-term creep performance was predicted by TTSP. Similarly, Jia et al. (2011) studied the effect of filler content and temperature on the creep and recovery behavior of multi-walled carbon nanotube (MWCNT)/polypropylene composites. TTSP was also applied to predict the long-term creep behavior. The creep strain was found to reduce with the increase in filler content and decrease in temperature. The creep behavior of alkali-treated jute reinforced PP composites (Chandekar and Chaudhari 2016) and starch grafted kenaf/PP composites (Hamma et al. 2014) was also explored by some researchers.

Xu et al. (2010) studied the creep behavior of bagasse fiber reinforced composites with virgin and recycled polyvinyl chloride (B/PVC), high density polyethylene (B/HDPE), and a commercial wood fiber/HDPE composite. The creep deformation of all of the composites was affected more significantly by the increase in temperature. The creep resistance of B/PVC was better than B/HDPE at low temperature, but they showed higher temperature dependence. The TTSP better predicted long-term creep behavior of the PVC composites than the HDPE composites. The effect of fiber size on the creep resistance of wood fiber/HDPE composites was explored in an other study (Wang et al. 2015). The large sized wood fiber composites exhibited better creep resistance at all temperatures as compared to short wood fiber composites. The creep deformation of fique/LDPE composites was decreased by treating reinforcing fique fibers with alkali, silane, and pre-impregnation with polyethylene due to increase in fiber/matrix interfacial adhesion (Hidalgo-Salazar et al. 2013). However, the effect of interfacial adhesion due to silane and pre-impregnation with polyethylene on the creep response of composites was better.

The creep behavior was investigated using different thermoset matrices including, MWCNT/epoxy (Starkova et al. 2012), biofiber face honeycomb core sandwich/epoxy (Du et al. 2013), kenaf/unsaturated polyester (Osman and Mutasher 2014), flax/vinylester (Amiri et al. 2015), and bark cloth/epoxy (Rwawiire et al. 2016) composites.

The effect of nanofiller type on the creep response of latex modified polyamide-6 nanocomposites was investigated by Siengchin and Kocsis (Siengchin and Karger-Kocsis 2009). The study revealed that the creep resistance of composites was dependent on the type of nanofiller mainly due to fundamental morphological differences. Similarly, some researchers also discovered the influence of nanofiller (Zhang et al. 2004), the effect of nanofiller size and type (Yang et al. 2006a) on the creep response of polyamide-66 nanocomposites.

The creep resistance of starch films by incorporating starch nanoparticles (Shi et al. 2013) and cellulose nanofibrils (CNFs) (Li et al. 2015) was found to increase, but CNFs concentration above 20% resulted in decrease in creep resistance due to poor dispersion of nanofibrils. TTSP was successfully used to predict the long-term creep performance of these films.

Yang et al. (2015) studied the creep behavior of bamboo fiber reinforced recycled PLA composites (BFRPCs) with fiber loading in the range of 0–80 wt%. The results exhibited the best creep resistance of BFRPC with 60 wt% fiber among all the composites. The effect of fiber type, stress, and temperature on the creep response of wood fiber reinforced biodegradable composites with two fiber loadings (20 wt% and 30 wt%) and containing matrix with a blend of poly(butylene adipate-terephthalate) (PBAT) copolyester and polylactic acid (PLA) was explored (Georgiopoulos et al. 2015). The creep deformation of composites showed a significant dependence on fiber type, temperature, and stress. The best creep resistance was presented by Lignocel® wood fibers at fiber loading of 20 wt%, probably due to the different nature of the fiber or/and its higher length.

2.12 Creep Models

Four parameters (or the Burger's) model is one of the mostly used physical models to give the relationship between the morphology of polymer composites and their creep behavior (Findley et al. 1989; Ward and Sweeney 2012). It is based on a series combination of a Maxwell element with a Kelvin–Voigt element as shown in Fig. 2.21 (Yang et al. 2006b). The total creep strain is divided into three separate parts: ε_M the instantaneous elastic deformation (Maxwell spring), ε_K viscoelastic deformation (Kelvin unit), and ε_∞ viscous deformation (Maxwell dash-pot). Thus, total strain as a function of time can be represented by Eqs. 2.2 and 2.3:

$$\varepsilon(t) = \varepsilon_M + \varepsilon_K + \varepsilon_\infty, \qquad (2.2)$$

Fig. 2.21 A schematic
representation of Burger's
model (Yang et al. 2006b)

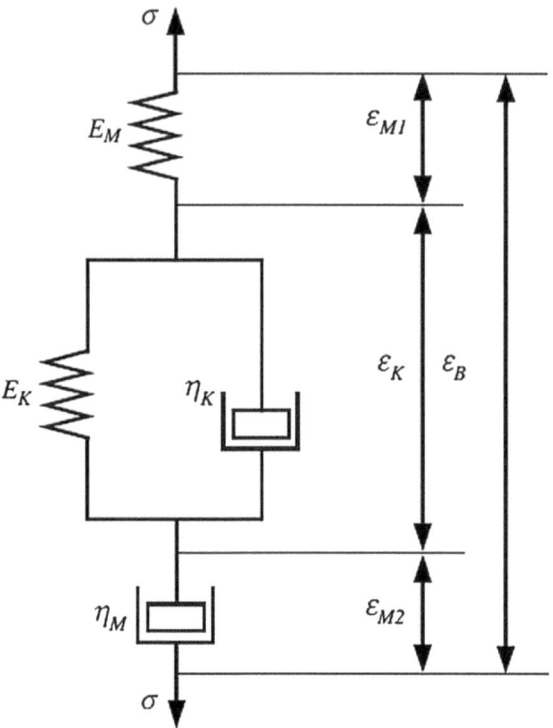

Fig. 2.21 A schematic representation of Burger's model (Yang et al. 2006b)

$$\varepsilon(t) = \frac{\sigma_0}{E_M} + \frac{\sigma_0}{E_K}\left(1 - e^{-E_K t/\eta_K}\right) + \frac{\sigma_0}{\eta_M}t, \qquad (2.3)$$

where $\varepsilon(t)$ is the creep strain, σ_0 is the stress, t is the time, E_M and E_K are the elastic moduli of Maxwell and Kelvin springs, and η_M and η_K are the viscosities of Maxwell and Kelvin dashpots. η_K/E_K is usually denoted as τ, the retardation time required to generate 63.2% deformation in the Kelvin unit (Yang et al. 2006b). ε_M is a constant value and does not change with time. ε_K represents the earliest stage of creep and attains a saturation value in short time, and ε_∞ represents the trend in the creep strain at sufficiently long time and appears similar to the deformation of a viscous liquid obeying Newton's law of viscosity.

The four parameters E_M, E_K, η_M, η_K can be obtained by fitting Eq. 2.3 to the experimental data and can be used to describe the creep behavior of composites. The creep rate of viscoelastic materials can be obtained by taking the derivative of Eq. 2.3.

$$\frac{d\varepsilon(t)}{dt} = \frac{\sigma_0}{E_K}\left(e^{-E_K t/\eta_K}\right) + \frac{\sigma_0}{\eta_M}. \qquad (2.4)$$

The Findley's power law model is an empirical mathematical model used to simulate the creep behavior of polymer composites. The model can be represented Eq. 2.5 (Findley et al. 1989);

$$\varepsilon(t) = at^b + \varepsilon_0, \tag{2.5}$$

where a and b are the material constants and ε_0 is the instantaneous strain. The ability of Findley's power law model to simulate the creep data has been found to be satisfactory in several studies (Yang et al. 2006b; Plaseied and Fatemi 2008; Jia et al. 2011). However, this model is not able to explain the creep mechanism of material. A two parameter empirical power law model has also been used in some studies (Tajvidi et al. 2005; Xu et al. 2010) to simulate the creep date. It has the form;

$$\varepsilon(t) = at^b, \tag{2.6}$$

where a and b are the material constants. The long-term creep is an important parameter to evaluate the end-use performance of composites, but it is often impractical to perform a creep test for an extremely long period of time. Time-temperature superposition principle (TTSP) is one of the common estimation techniques to predict the long-term creep behavior by shifting the curves from tests at different temperatures horizontally along the logarithmic time axis to generate a single curve known as master curve (Ward and Sweeney 2012). The shifting distance is called shift factor. According to this principle, the viscoelastic response of a material at a higher temperature is identical with the response of same material at the low temperature for a longer time.

References

Abdallah, M., Zitoune, R., Collombet, F., & Bezzazi, B. (2010). Study of mechanical and thermomechanical properties of jute/epoxy composite laminate. *Journal of Reinforced Plastics and Composites, 29*(11), 1669–1680.

Acha, B. A., Reboredo, M. M., & Marcovich, N. E. (2007). Creep and dynamic mechanical behavior of PP–jute composites: Effect of the interfacial adhesion. *Composites Part A: Applied Science and Manufacturing, 38*(6), 1507–1516.

Akovali, G. (2001) *Handbook of composite fabrication.* Shropshire: iSmithers Rapra Publishing.

Al-Qureshi, H. (2001) The application of jute fibre reinforced composites for the development of a car body. In *UMIST Conference, UK.*

Alves, C., Silva, A., Reis, L., Freitas, M., Rodrigues, L., & Alves, D. (2010). Ecodesign of automotive components making use of natural jute fiber composites. *Journal of Cleaner Production, 18*(4), 313–327.

Amiri, A., Hosseini, N., & Ulven, C. A. (2015). Long-term creep behavior of flax/vinyl ester composites using time-temperature superposition principle. *Journal of Renewable Materials, 3*(3), 224–233.

Araujo, J., Waldman, W., & De Paoli, M. (2008). Thermal properties of high density polyethylene composites with natural fibres: Coupling agent effect. *Polymer Degradation and Stability, 93* (10), 1770–1775.

Avancha, S., Behera, A. K., Sen, R., & Adhikari, B. (2013). Physical and mechanical characterization of jute reinforced soy composites. *Journal of Reinforced Plastics and Composites, 32*(18), 1380–1390.

Azizi Samir, M. A. S., Alloin, F., & Dufresne, A. (2005). Review of recent research into cellulosic whiskers, their properties and their application in nanocomposite field. *Biomacromolecules, 6* (2), 612–626.

Azwa, Z., Yousif, B., Manalo, A., & Karunasena, W. (2013). A review on the degradability of polymeric composites based on natural fibres. *Materials and Design, 47,* 424–442.

Baltazar-y-Jimenez, A., Bistritz, M., Schulz, E., & Bismarck, A. (2008). Atmospheric air pressure plasma treatment of lignocellulosic fibres: Impact on mechanical properties and adhesion to cellulose acetate butyrate. *Composites Science and Technology, 68*(1), 215–227.

Betten, J. (2008) Creep behavior of isotropic and anisotropic materials; Constitutive equations. In *Creep mechanics* (pp. 49–83). Berlin: Springer.

Biswas, B., Chabri, S., Sawai, P., Mitra, B. C., Das, K., & Sinha, A. (2016). Effect of copper incorporation on the mechanical and thermal behavior of jute fiber reinforced unsaturated polyester composites. *Polymer Composites.* doi:10.1002/pc.24181.

Carvelli, V., Betti, A., & Fujii, T. (2016). Fatigue and Izod impact performance of carbon plain weave textile reinforced epoxy modified with cellulose microfibrils and rubber nanoparticles. *Composites Part A: Applied Science and Manufacturing, 84,* 26–35.

Chandekar, H., & Chaudhari, V. (2016) Flexural creep behaviour of jute polypropylene composites. In *IOP Conference Series: Materials Science and Engineering.* IOP Publishing. doi:10.1088/1757-899X/149/1/012107.

Chandramohan, D., & Marimuthu, K. (2011). A review on natural fibers. *International Journal of Research and Reviews in Applied Sciences, 8*(2), 194–206.

Chirayil, C. J., Joy, J., Mathew, L., Koetz, J., & Thomas, S. (2014). Nanofibril reinforced unsaturated polyester nanocomposites: morphology, mechanical and barrier properties, viscoelastic behavior and polymer chain confinement. *Industrial Crops and Products, 56,* 246–254.

Cho, M.-J., & Park, B.-D. (2011). Tensile and thermal properties of nanocellulose-reinforced poly (vinyl alcohol) nanocomposites. *Journal of Industrial and Engineering Chemistry, 17*(1), 36–40.

Dash, B., Rana, A., Mishra, H., Nayak, S., & Tripathy, S. (2000a). Novel low-cost jute–polyester composites. III. Weathering and thermal behavior. *Journal of Applied Polymer Science, 78*(9), 1671–1679.

Dash, B., Rana, A., Mishra, S., Mishra, H., Nayak, S., & Tripathy, S. (2000b). Novel low-cost jute–polyester composite. II. SEM observation of the fractured surfaces. *Polymer-Plastics Technology and Engineering, 39*(2), 333–350.

Dittenber, D. B., & GangaRao, H. V. (2012). Critical review of recent publications on use of natural composites in infrastructure. *Composites Part A: Applied Science and Manufacturing, 43*(8), 1419–1429.

Du, Y., Yan, N., & Kortschot, M. T. (2013). An experimental study of creep behavior of lightweight natural fiber-reinforced polymer composite/honeycomb core sandwich panels. *Composite Structures, 106,* 160–166.

Dwivedi, U., & Chand, N. (2009). Influence of fibre orientation on friction and sliding wear behaviour of jute fibre reinforced polyester composite. *Applied Composite Materials, 16*(2), 93–100.

Faruk, O., Bledzki, A. K., Fink, H.-P., & Sain, M. (2012). Biocomposites reinforced with natural fibers: 2000–2010. *Progress in Polymer Science, 37*(11), 1552–1596.

Fengel, D., & Wegener, G. (1983) *Wood: Chemistry, ultrastructure, reactions.* New York: Walter de Gruyter.

Findley, W. N., Davis, F. A., & Onaran, K. (1989). *Creep and relaxation of nonlinear viscoelastic materials: With an introduction to linear viscoelasticity.* New York: Dover Publications Inc.

Fraga, A., Frullloni, E., De la Osa, O., Kenny, J., & Vázquez, A. (2006). Relationship between water absorption and dielectric behaviour of natural fibre composite materials. *Polymer Testing, 25*(2), 181–187.

Gao, S.-L., & Mäder, E. (2006). Jute/polypropylene composites I. Effect of matrix modification. *Composites Science and Technology, 66*(7), 952–963.

Gassan, J., & Bledzki, A. K. (1997). The influence of fiber-surface treatment on the mechanical properties of jute-polypropylene composites. *Composites Part A: Applied Science and Manufacturing, 28*(12), 1001–1005.

Gassan, J., & Bledzki, A. K. (1999). Possibilities for improving the mechanical properties of jute/epoxy composites by alkali treatment of fibres. *Composites Science and Technology, 59* (9), 1303–1309.

George, J., Sreekala, M., & Thomas, S. (2001). A review on interface modification and characterization of natural fiber reinforced plastic composites. *Polymer Engineering & Science, 41*(9), 1471–1485.

Georgiopoulos, P., Kontou, E., & Christopoulos, A. (2015). Short-term creep behavior of a biodegradable polymer reinforced with wood-fibers. *Composites Part B: Engineering, 80,* 134–144.

Goriparthi, B. K., Suman, K., & Rao, N. M. (2012). Effect of fiber surface treatments on mechanical and abrasive wear performance of polylactide/jute composites. *Composites Part A: Applied Science and Manufacturing, 43*(10), 1800–1808.

Gurunathan, T., Mohanty, S., & Nayak, S. K. (2015). A review of the recent developments in biocomposites based on natural fibres and their application perspectives. *Composites Part A: Applied Science and Manufacturing, 77,* 1–25.

Hamma, A., Kaci, M., Ishak, Z. M., & Pegoretti, A. (2014). Starch-grafted-polypropylene/kenaf fibres composites. Part 1: Mechanical performances and viscoelastic behaviour. *Composites Part A: Applied Science and Manufacturing, 56,* 328–335.

Hao, A., Chen, Y., & Chen, J. Y. (2014) Creep and recovery behavior of kenaf/polypropylene nonwoven composites. *Journal of applied polymer science, 131*(17), doi:10.1002/app.40726.

Helbert, W., Cavaille, J., & Dufresne, A. (1996). Thermoplastic nanocomposites filled with wheat straw cellulose whiskers. Part I: Processing and mechanical behavior. *Polymer Composites, 17* (4), 604–611.

Hidalgo-Salazar, M. A., Mina, J. H., & Herrera-Franco, P. J. (2013). The effect of interfacial adhesion on the creep behaviour of LDPE–Al–Fique composite materials. *Composites Part B: Engineering, 55,* 345–351.

Hill, S. (1997). Cars that grow on trees. *New Scientists, 2067,* 36–39.

Hobson, J., & Carus, M. (2011). Targets for bio-based composites and natural fibres. *JEC composites, 63,* 31–32.

Hong, C., Hwang, I., Kim, N., Park, D., Hwang, B., & Nah, C. (2008). Mechanical properties of silanized jute–polypropylene composites. *Journal of Industrial and Engineering Chemistry, 14* (1), 71–76.

Jafari, M., Nikkhah, A., Sadeghi, A., & Chamani, M. (2007). The effect of Pleurotus spp. fungi on chemical composition and in vitro digestibility of rice straw. *Pakistan Journal of Biological Sciences, 10*(15), 2460–2464.

Jawaid, M., & Khalil, H. A. (2011). Cellulosic/synthetic fibre reinforced polymer hybrid composites: A review. *Carbohydrate Polymers, 86*(1), 1–18.

Jia, Y., Peng, K., Gong, X.-L., & Zhang, Z. (2011). Creep and recovery of polypropylene/carbon nanotube composites. *International Journal of Plasticity, 27*(8), 1239–1251.

Jiang, B., Liu, C., Zhang, C., Wang, B., & Wang, Z. (2007). The effect of non-symmetric distribution of fiber orientation and aspect ratio on elastic properties of composites. *Composites Part B Engineering, 38*(1), 24–34.

John, M. J., & Anandjiwala, R. D. (2008). Recent developments in chemical modification and characterization of natural fiber-reinforced composites. *Polymer Composites, 29*(2), 187–207.

Jonoobi, M., Harun, J., Mathew, A. P., & Oksman, K. (2010). Mechanical properties of cellulose nanofiber (CNF) reinforced polylactic acid (PLA) prepared by twin screw extrusion. *Composites Science and Technology, 70*(12), 1742–1747.

Jonoobi, M., Niska, K. O., Harun, J., & Misra, M. (2009). Chemical composition, crystallinity, and thermal degradation of bleached and unbleached kenaf bast (Hibiscus cannabinus) pulp and nanofibers. *BioResources, 4*(2), 626–639.

Joseph, P., Joseph, K., Thomas, S., Pillai, C., Prasad, V., Groeninckx, G., et al. (2003). The thermal and crystallisation studies of short sisal fibre reinforced polypropylene composites. *Composites Part A: Applied Science and Manufacturing, 34*(3), 253–266.

Kabir, M., Wang, H., Lau, K., & Cardona, F. (2012). Chemical treatments on plant-based natural fibre reinforced polymer composites: An overview. *Composites Part B: Engineering, 43*(7), 2883–2892.

Kalia, S., Kaith, B., & Kaur, I. (2009). Pretreatments of natural fibers and their application as reinforcing material in polymer composites—A review. *Polymer Engineering & Science, 49*(7), 1253–1272.

Kalia, S., Thakur, K., Celli, A., Kiechel, M. A., & Schauer, C. L. (2013). Surface modification of plant fibers using environment friendly methods for their application in polymer composites, textile industry and antimicrobial activities: A review. *Journal of Environmental Chemical Engineering, 1*(3), 97–112.

Karaduman, Y., & Onal, L. (2013). Dynamic mechanical and thermal properties of enzyme-treated jute/polyester composites. *Journal of Composite Materials, 47*(19), 2361–2370.

Keener, T., Stuart, R., & Brown, T. (2004). Maleated coupling agents for natural fibre composites. *Composites Part A: Applied Science and Manufacturing, 35*(3), 357–362.

Khan, J. A., Khan, M. A., & Islam, R. (2013). Mechanical, thermal and degradation properties of jute fabric–reinforced polypropylene composites: Effect of potassium permanganate as oxidizing agent. *Polymer Composites, 34*(5), 671–680.

Khan, R. A., Khan, M. A., Khan, A., & Hossain, M. (2009). Effect of gamma radiation on the performance of jute fabrics-reinforced polypropylene composites. *Radiation Physics and Chemistry, 78*(11), 986–993.

Kicińska-Jakubowska, A., Bogacz, E., & Zimniewska, M. (2012). Review of natural fibers. Part I—vegetable fibers. *Journal of Natural Fibers, 9*(3), 150–167.

Kvien, I., & Oksman, K. (2007). Orientation of cellulose nanowhiskers in polyvinyl alcohol. *Applied Physics A, 87*(4), 641–643.

Lee, H., Hamid, S., & Zain, S. (2014). Conversion of lignocellulosic biomass to nanocellulose: Structure and chemical process. *The Scientific World Journal, 2014*, 20.

Lee, K.-Y., Delille, A., & Bismarck, A. (2011). Greener surface treatments of natural fibres for the production of renewable composite materials. In *Cellulose fibers: Bio-and nano-polymer composites* (pp. 155–178). Berlin: Springer.

Li, M., Li, D., Wang, L.-J., & Adhikari, B. (2015). Creep behavior of starch-based nanocomposite films with cellulose nanofibrils. *Carbohydrate Polymers, 117*, 957–963.

Ma, H., & Joo, C. W. (2011). Structure and mechanical properties of jute-polylactic acid biodegradable composites. *Journal of Composite Materials, 45*(14), 1451–1460.

Mohanty, A., Khan, M., & Hinrichsen, G. (2000a). Influence of chemical surface modification on the properties of biodegradable jute fabrics—polyester amide composites. *Composites Part A: Applied Science and Manufacturing, 31*(2), 143–150.

Mohanty, A., Khan, M. A., & Hinrichsen, G. (2000b). Surface modification of jute and its influence on performance of biodegradable jute-fabric/Biopol composites. *Composites Science and Technology, 60*(7), 1115–1124.

Mohanty, A., Misra, M., & Hinrichsen, G. (2000c). Biofibres, biodegradable polymers and biocomposites: An overview. *Macromolecular Materials and Engineering, 276*(1), 1–24.

Mohanty, A. K., Misra, M., & Drzal, L. T. (2005) *Natural fibers, biopolymers, and biocomposites*. Boca Raton: CRC Press.

Mwaikambo, L. Y., & Ansell, M. P. (2002). Chemical modification of hemp, sisal, jute, and kapok fibers by alkalization. *Journal of Applied Polymer Science, 84*(12), 2222–2234.

Ng, H.-M., Sin, L. T., Tee, T.-T., Bee, S.-T., Hui, D., Low, C.-Y., et al. (2015). Extraction of cellulose nanocrystals from plant sources for application as reinforcing agent in polymers. *Composites Part B: Engineering, 75,* 176–200.

Novosel'skaya, I., Voropaeva, N., Semenova, L., & Rashidova, S. S. (2000). Trends in the science and applications of pectins. *Chemistry of Natural Compounds, 36*(1), 1–10.

Nuñez, A. J., Marcovich, N. E., & Aranguren, M. I. (2004). Analysis of the creep behavior of polypropylene-woodflour composites. *Polymer Engineering & Science, 44*(8), 1594–1603.

Osman, E. A. and Mutasher, S. A. (2014). Viscoelastic properties of kenaf reinforced unsaturated polyester composites. *International Journal of Computational Materials Science and Engineering, 3*(01), doi:10.1142/S2047684114500043.

Pallesen, B. E. (1996). The quality of combine-harvested fibre flax for industrials purposes depends on the degree of retting. *Industrial Crops and Products, 5*(1), 65–78.

Pan, N., Day, A., & Mahalanabis, K. (2000). Properties of jute. *Indian Textile Journal, 110*(5), 16–23.

Peijs, T. (2003). Composites for recyclability. *Materials Today, 6*(4), 30–35.

Pickering, K. (2008) *Properties and performance of natural-fibre composites.* Amsterdam: Elsevier.

Pickering, K. L., Li, Y., Farrell, R. L., & Lay, M. (2007). Interfacial modification of hemp fiber reinforced composites using fungal and alkali treatment. *Journal of Biobased Materials and Bioenergy, 1*(1), 109–117.

Piggott, M. (1980) *Load bearing fibre composites.* International series on the strength and fracture of material and structures. Oxford: Pergamon Press.

Plackett, D., Andersen, T. L., Pedersen, W. B., & Nielsen, L. (2003). Biodegradable composites based on L-polylactide and jute fibres. *Composites Science and Technology, 63*(9), 1287–1296.

Plaseied, A., & Fatemi, A. (2008). Tensile creep and deformation modeling of vinyl ester polymer and its nanocomposite. *Journal of Reinforced Plastics and Composites, 28*(14), 1775–1788.

Ragoubi, M., Bienaimé, D., Molina, S., George, B., & Merlin, A. (2010). Impact of corona treated hemp fibres onto mechanical properties of polypropylene composites made thereof. *Industrial Crops and Products, 31*(2), 344–349.

Rahman, M. M., Mallik, A. K., & Khan, M. A. (2007). Influences of various surface pretreatments on the mechanical and degradable properties of photografted oil palm fibers. *Journal of Applied Polymer Science, 105*(5), 3077–3086.

Ranganathan, S. R. and Quayyum, Z. (1993). *New horizons for jute.* Ahmedabad: National Information Centre for Textile and Allied Subjects.

Ray, D., Sarkar, B., Basak, R., & Rana, A. (2004). Thermal behavior of vinyl ester resin matrix composites reinforced with alkali-treated jute fibers. *Journal of Applied Polymer Science, 94* (1), 123–129.

Ray, D., Sarkar, B., Das, S., & Rana, A. (2002a). Dynamic mechanical and thermal analysis of vinylester-resin-matrix composites reinforced with untreated and alkali-treated jute fibres. *Composites Science and Technology, 62*(7), 911–917.

Ray, D., Sarkar, B. K., & Bose, N. R. (2002b). Impact fatigue behaviour of vinylester resin matrix composites reinforced with alkali treated jute fibres. *Composites Part A: Applied Science and Manufacturing, 33*(2), 233–241.

Rojas, J., Bedoya, M. and Ciro, Y. (2015) Current trends in the production of cellulose nanoparticles and nanocomposites for biomedical applications. In M. Poletto & H. L. Ornaghi (Eds.), *Cellulose—fundamental aspects and current trends.* Rijeka: InTech.

Rong, M. Z., Zhang, M. Q., Liu, Y., Yang, G. C., & Zeng, H. M. (2001). The effect of fiber treatment on the mechanical properties of unidirectional sisal-reinforced epoxy composites. *Composites Science and Technology, 61*(10), 1437–1447.

Rwawiire, S., Tomkova, B., Wiener, J., Militky, J., Kasedde, A., Kale, B. M., et al. (2016). Short-term creep of barkcloth reinforced laminar epoxy composites. *Composites Part B Engineering, 103,* 131–138.

Saheb, D. N., & Jog, J. (1999). Natural fiber polymer composites: A review. *Advances in Polymer Technology, 18*(4), 351–363.

Santulli, C. (2001). Post-impact damage characterisation on natural fibre reinforced composites using acoustic emission. *NDT and E International, 34*(8), 531–536.

Sarikanat, M. (2009). The influence of oligomeric siloxane concentration on the mechanical behaviors of alkalized jute/modified epoxy composites. *Journal of Reinforced Plastics and Composites, 29*(6), 807–817.

Sarkar, B., & Ray, D. (2004). Effect of the defect concentration on the impact fatigue endurance of untreated and alkali treated jute–vinylester composites under normal and liquid nitrogen atmosphere. *Composites Science and Technology, 64*(13), 2213–2219.

Sarkar, S., & Adhikari, B. (2001). Jute felt composite from lignin modified phenolic resin. *Polymer Composites, 22*(4), 518–527.

Satyanarayana, K. (2004). Steam explosion—a boon for value addition to renewable resources. *Metal News, 22*, 35–40.

Satyanarayana, K., Ravikumar, K., Sukumaran, K., Mukherjee, P., Pillai, S., & Kulkarni, A. (1986). Structure and properties of some vegetable fibres. *Journal of Materials Science, 21*(1), 57–63.

Sedan, D., Pagnoux, C., Smith, A., & Chotard, T. (2008). Mechanical properties of hemp fibre reinforced cement: Influence of the fibre/matrix interaction. *Journal of the European Ceramic Society, 28*(1), 183–192.

Sever, K., Sarikanat, M., Seki, Y., Erkan, G., & Erdoğan, Ü. H. (2010). The mechanical properties of γ-methacryloxypropyltrimethoxy silane-treated jute/polyester composites. *Journal of Composite Materials, 44*(15), 1913–1924.

Shi, A.-M., Wang, L.-J., Li, D., & Adhikari, B. (2013). Characterization of starch films containing starch nanoparticles. Part 2: Viscoelasticity and creep properties. *Carbohydrate Polymers, 96*(2), 602–610.

Siengchin, S., & Karger-Kocsis, J. (2009). Structure and creep response of toughened and nanoreinforced polyamides produced via the latex route: Effect of nanofiller type. *Composites Science and Technology, 69*(5), 677–683.

Singh, B., Gupta, M., & Verma, A. (2000). The durability of jute fibre-reinforced phenolic composites. *Composites Science and Technology, 60*(4), 581–589.

Siró, I., & Plackett, D. (2010). Microfibrillated cellulose and new nanocomposite materials: A review. *Cellulose, 17*(3), 459–494.

Sreekala, M., Kumaran, M., Joseph, S., Jacob, M., & Thomas, S. (2000). Oil palm fibre reinforced phenol formaldehyde composites: Influence of fibre surface modifications on the mechanical performance. *Applied Composite Materials, 7*(5–6), 295–329.

Starkova, O., Buschhorn, S., Mannov, E., Schulte, K., & Aniskevich, A. (2012). Creep and recovery of epoxy/MWCNT nanocomposites. *Composites Part A: Applied Science and Manufacturing, 43*(8), 1212–1218.

Suddell, B., & Evans, W. (2003). The increasing use and application of natural fiber composite materials within the automotive industry. In *Seventh Composite Conference on Woodfiber–Plastic Composites*, 7–14.

Sudha, S., & Thilagavathi, G. (2016). Effect of alkali treatment on mechanical properties of woven jute composites. *The Journal of The Textile Institute, 107*(6), 691–701.

Tajvidi, M., Falk, R. H., & Hermanson, J. C. (2005). Time–temperature superposition principle applied to a kenaf-fiber/high-density polyethylene composite. *Journal of Applied Polymer Science, 97*(5), 1995–2004.

Tao, Y., Yan, L., & Jie, R. (2009). Preparation and properties of short natural fiber reinforced poly (lactic acid) composites. *Transactions of Nonferrous Metals Society of China, 19*, s651–s655.

Tsoumis, G. (1991) *Science and technology of wood: Structure, properties, utilization*. New York: Van Nostrand Reinhold.

Van de Weyenberg, I., Truong, T. C., Vangrimde, B., & Verpoest, I. (2006). Improving the properties of UD flax fibre reinforced composites by applying an alkaline fibre treatment. *Composites Part A Applied Science and Manufacturing, 37*(9), 1368–1376.

Wang, W. H., Huang, H. B., Du, H. H., & Wang, H. (2015). Effects of fiber size on short-term creep behavior of wood fiber/HDPE composites. *Polymer Engineering & Science, 55*(3), 693–700.

Wang, X., Cui, Y., Xu, Q., Xie, B., & Li, W. (2010). Effects of alkali and silane treatment on the mechanical properties of jute-fiber-reinforced recycled polypropylene composites. *Journal of Vinyl and Additive Technology, 16*(3), 183–188.

Ward, I. M., & Sweeney, J. (2012). *Mechanical properties of solid polymers*. New Jersey: Wiley.

Xu, S., Girouard, N., Schueneman, G., Shofner, M. L., & Meredith, J. C. (2013). Mechanical and thermal properties of waterborne epoxy composites containing cellulose nanocrystals. *Polymer, 54*(24), 6589–6598.

Xu, Y., Wu, Q., Lei, Y., & Yao, F. (2010). Creep behavior of bagasse fiber reinforced polymer composites. *Bioresource Technology, 101*(9), 3280–3286.

Yang, J.-L., Zhang, Z., Schlarb, A. K., & Friedrich, K. (2006a). On the characterization of tensile creep resistance of polyamide 66 nanocomposites. Part I. Experimental results and general discussions. *Polymer, 47*(8), 2791–2801.

Yang, J.-L., Zhang, Z., Schlarb, A. K., & Friedrich, K. (2006b). On the characterization of tensile creep resistance of polyamide 66 nanocomposites. Part II: Modeling and prediction of long-term performance. *Polymer, 47*(19), 6745–6758.

Yang, T.-C., Wu, T.-L., Hung, K.-C., Chen, Y.-L., & Wu, J.-H. (2015). Mechanical properties and extended creep behavior of bamboo fiber reinforced recycled poly (lactic acid) composites using the time–temperature superposition principle. *Construction and Building Materials, 93*, 558–563.

Zaman, H. U., Khan, R. A., Haque, M., Khan, M. A., Khan, A., Huq, T., et al. (2010). Preparation and mechanical characterization of jute reinforced polypropylene/natural rubber composite. *Journal of Reinforced Plastics and Composites, 29*(20), 3064–3065.

Zhang, Z., Yang, J.-L., & Friedrich, K. (2004). Creep resistant polymeric nanocomposites. *Polymer, 45*(10), 3481–3485.

Zimniewska, M., Wladyka-Przybylak, M., & Mankowski, J. (2011). Cellulosic bast fibers, their structure and properties suitable for composite applications. In *Cellulose fibers: Bio-and nano-polymer composite* (pp. 97–119). Berlin: Springer.

Chapter 3
Research Methodology

Abstract This chapter gives information about the materials used to prepare jute/green epoxy composites. The methods used to prepare nanocellulose and pulverized jute fibers (PJF) from waste jute are elaborated. Different novel fiber treatment methods for surface treatment are introduced. The chapter further goes with different techniques for characterization of fibers, i.e., scanning electron microscopy (SEM), Fourier transform infrared spectroscopy (FTIR), and X-ray diffraction (XRD). After that, standards adopted for mechanical and thermomechanical testing of prepared composites are described and the way to fit the creep data with different creep models is presented.

Keywords Surface treatments · Hand layup · Mechanical testing · Creep modeling

3.1 Materials

Jute yarn having linear density 386 tex and 124 twists/m, produced from tossa jute (*C. olitorius*) fibers, was used to produce a woven fabric as shown in Fig. 3.1 having areal density of 600 gm^{-2} with 5-end satin weave design on a shuttle loom. Warp and weft densities of the fabric were 6.3 threads per cm and 7.9 threads per cm, respectively. Jute fabric was washed with 2 wt% non-ionic detergent solution at 70 °C for 30 min to remove any dirt and impurities and dried at room temperature for 48 h.

Waste jute fibers, sourced from a jute mill, were used for pulverization and purification and extraction of nanocellulose. Green epoxy resin CHS-Epoxy G520 and hardener TELALIT 0600 were supplied by Spolchemie, Czech Republic. The main characteristics of the resin system are reported in Table 3.1 as specified by the manufacturer. Sulfuric acid (H_2SO_4) and sodium hydroxide (NaOH) were supplied by Lach-Ner, Czech Republic. Sodium sulfate (Na_2SO_4) and sodium hypochlorite (NaOCl) were supplied by Sigma-Aldrich, Czech Republic.

© The Author(s) 2017

A. Jabbar, *Sustainable Jute-Based Composite Materials*, SpringerBriefs in Applied Sciences and Technology, DOI 10.1007/978-3-319-65457-7_3

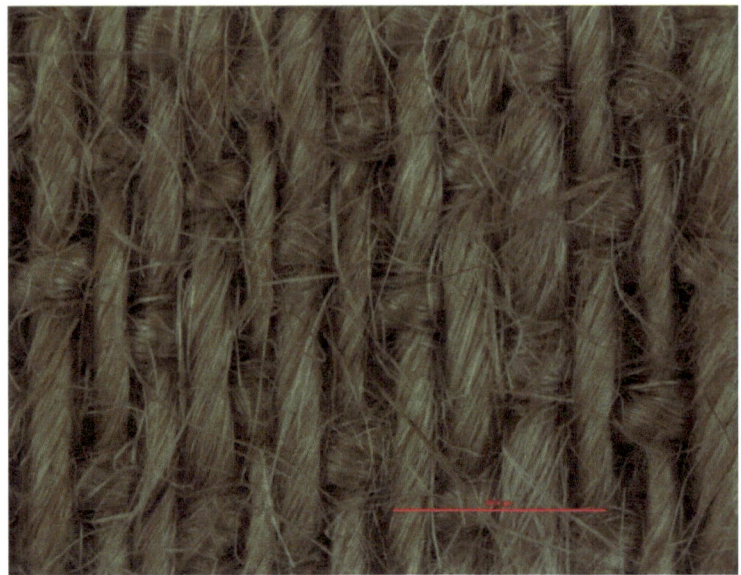

Fig. 3.1 Jute woven fabric used in the study

Table 3.1 Resin system characteristics

Characteristics	Resin + Hardener (CHS-Epoxy G520 + TELALIT 0600)
Viscosity (Pa.s, 25 °C)	3.8
Mixing ratio (pbw)	100:32
Minimal curing temperature (°C)	20
Minimal pot life (23 °C, h)	6
T_g (°C)	200
Flexural strength (MPa)	115
Tensile strength (MPa)	65
Elongation (%)	4
Impact strength (kJ/m^2)	20

3.2 Methodology

3.2.1 Chemical Pretreatment of Jute Fabric and Waste Jute Fibers

Figure 3.2 presents the flowchart of chemical pretreatment of jute fabric and waste jute fibers. The jute fabric and waste jute fibers were immersed separately in 2% NaOH solution for 1 h at 80 °C maintaining a liquor ratio of 15:1. Alkali-treated waste jute fibers were further treated with 7 g/l NaOCl solution at room temperature

Fig. 3.2 Process flowchart of chemical pretreatment of **a** jute fabric and **b** waste jute fibers

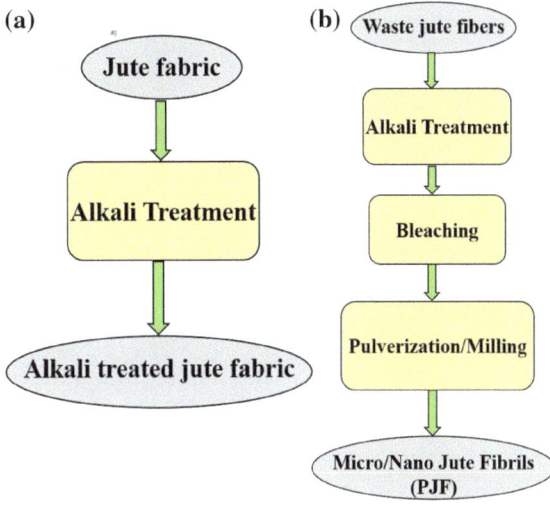

for 2 h under pH 10–11 and subsequently antichlored with 0.1% Na_2SO_4 at 50 °C for 20 min. Both fabric and waste fibers, after chemical pretreatment, were washed with freshwater several times until the final pH was maintained at 7.0 and then allowed to dry at room temperature for 48 h and at 100 °C in an oven for 2 h.

3.2.2   Pulverization of Waste Jute Fibers

Pulverization of chemically treated waste jute fibers was carried out using a high-energy planetary ball mill of Fritsch pulverisette 7. Pulverization process relies on the principle of energy release at the point of impact between balls and on the high grinding action created by friction of balls on the wall (Baheti and Militky 2013). The sintered corundum container of 80 ml capacity and zirconium balls of 10 mm diameter were chosen for 1.0 h of pulverization. The ball-to-material ratio (BMR) was kept at 10:1, and the speed was kept at 850 rpm.

3.2.3   Purification and Extraction of Cellulose from Waste Jute Fibers and Nanocellulose Coating

The main steps followed to purify and extract nanocellulose from waste jute fibers are depicted in Fig. 3.3. Waste jute fibers were chopped to approximate length of 5–10 mm and immersed in 2% sodium hydroxide (NaOH) solution for 2 h at 80 °C temperature maintaining a liquor ratio of 50:1. The process was repeated three times. The fibers were then washed with tap water several times to remove any traces of

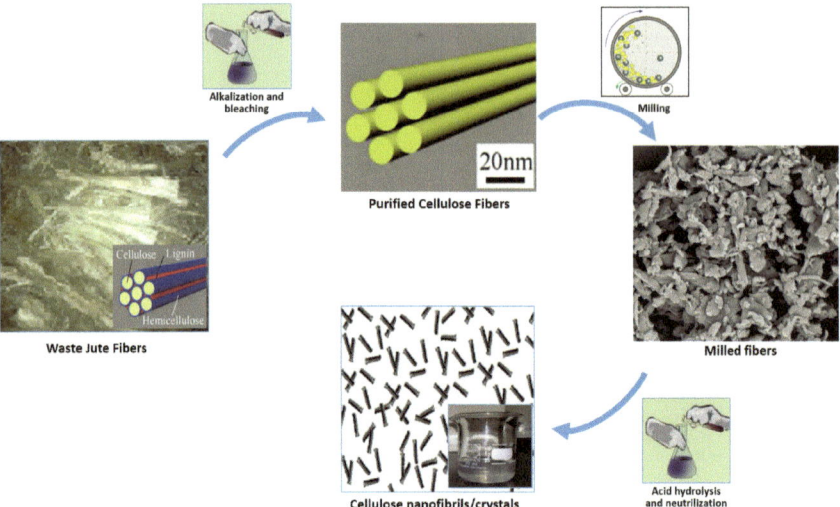

Fig. 3.3 Steps adopted in the purification and extraction of cellulose nanofibrils from waste jute fibers

NaOH sticking to the fibers' surface. The jute fibers were then bleached with 7 g/l sodium hypochlorite (NaOCl) solution at room temperature for 2 h under pH 10–11 and subsequently antichlored with 0.1% sodium sulfate (Na$_2$SO$_4$) at 50 °C for 20 min. Finally, the fibers were washed with tap water several times until the final pH was maintained at 7.0 and then allowed to dry at room temperature for 48 h and at 100 °C in an oven for 2 h. Alkali and bleaching treatments were used to purify cellulose and to remove maximum amount of hemicellulose and lignin from the fibers. The bleached jute fibers were milled using a high-energy planetary ball mill of Fritsch pulverisette 7. Milling process relies on the principle of energy release at the point of impact between balls and on the high grinding action created by friction of balls on the wall (Baheti and Militky 2013). The sintered corundum container of 80 ml capacity and zirconium balls of 10 mm diameter were chosen for 20 min of milling. The ball-to-material ratio (BMR) was kept at 10:1, and the speed was kept at 850 rpm. Acid hydrolysis of milled jute fibers was conducted for 1 h at 45 °C under mechanical stirring using 65% (w/w) H$_2$SO$_4$. The fiber content during acid hydrolysis was 5% (w/w). The suspension was diluted with cold water (4 °C) to stop the reaction, neutralized with NaOH solution, and discolored by NaOCl solution. The supernatant was removed from the sediment and replaced by new distilled water several times. Finally, 3, 5, and 10% (w/w) nanocellulose suspensions were prepared by increasing cellulose concentration and decreasing water concentration through filtration using Buchner funnel. The prepared nanocellulose suspensions (3, 5, and 10 wt%) were ultrasonicated for 5 min with Bandelin ultrasonic probe and then applied on the surface of jute fabric by roller padding at room temperature. Finally, the coated fabrics were dried at 70 °C for 60 min.

3.2.4   Treatment Methods

3.2.4.1   Enzyme Treatment

Untreated jute fabric was subjected to enzyme treatment. A solution, having 1% owf Texazym DLG new, 3% owf Texzym BFE, and 0.2 g/l of Texawet DAF anti-foaming agent (all supplied by INOTEX, Czech Republic) in distilled water, was prepared. Texazym DLG new catalyzes the decomposition of hemicellulose and partially lignin and can affect cellulose in fiber. Texzym BFE helps the removal and decomposition of interfiber binding substances. Jute fabric was dipped in the solution at 50 °C for 2 h maintaining a liquor ratio of 10:1. After the treatment, the fabric was rinsed with freshwater several times and dried at room temperature for 48 h.

3.2.4.2   Ozone Treatment

Ozone treatment was done by putting the jute fabric for 1 h in a closed container filled with ozone gas. The container was connected to ozone generator "TRIOTECH GO 5LAB-K" (TRIOTECH s.r.o. Czech Republic) which was continuously generating ozone gas at the rate of 5.0 g/h. Oxygen for the production of ozone gas was generated by "Kröber O2" (KröberMedizintechnik GmbH, Germany). The complete setup for generating ozone gas is shown in Fig. 3.4.

Fig. 3.4 Setup for generation of ozone gas

3.2.4.3 Laser Treatment

Laser irradiation was performed on the surface of jute fabric with a commercial carbon dioxide pulsed infrared (IR) laser "Marcatex 150 Flexi Easy-Laser" (Garment Finish Kay, S.L. Spain) as presented in Fig. 3.5, generating laser beam with a wavelength of 10.6 μm. Parameters that determine marking intensity of laser are marking speed (bits/ms), duty cycle (%), and frequency (kHz). In this study, the marking speed was set to 200 bits/ms, the duty cycle (DC) to 50%, and frequency to 5 kHz. The used laser power was 100 W. Laser beams interact with fibers by local evaporation of material, thermal decomposition, or changing the surface roughness (Stepankova et al. 2010).

3.2.4.4 Plasma Treatment

Jute fabric was treated for 60 s with dielectric barrier discharge (DBD) plasma with discharge power of 190 W at atmospheric pressure using a laboratory device (Universal Plasma Reactor, model FB-460, Czech Republic), shown in Fig. 3.6.

3.2.5 Preparation of Composites

The composite laminates were prepared by hand layup method. The resin and hardener were mixed in a ratio of 100:32 (by weight) according to manufacturer

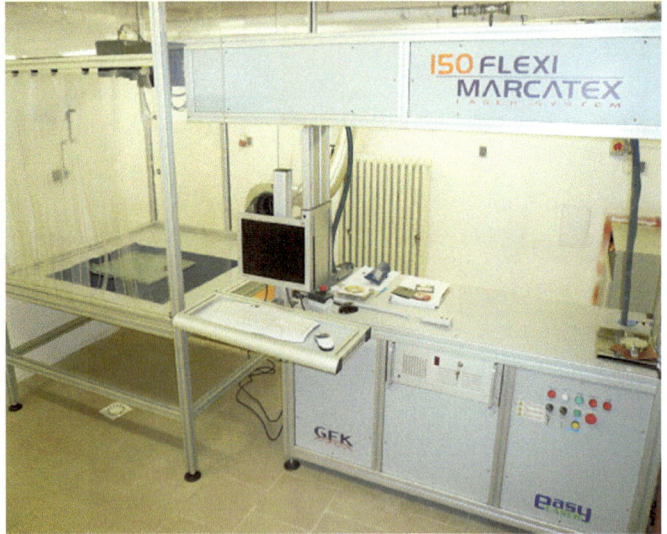

Fig. 3.5 Laser treatment setup

Fig. 3.6 Plasma device

recommendations, before hand layup. Three different categories of composite laminates were prepared using same green epoxy matrix. The first type of composite was consisted of 1, 5, and 10 wt% of pulverized micro jute fibers (PJF) used as fillers along with alkali-treated jute fabric. The second type was enclosed with nanocellulose-coated jute fabric with different cellulose concentrations (3, 5, and 10 wt%), and third type was comprised of novel surface-treated jute fabrics. The composite layup along with Teflon sheets were sandwiched between a pair of steel plates and cured at 120 °C for 1.0 h in mechanical convection oven with predetermined weight on it to maintain uniform pressure of about 50 kPa (Mishra et al. 2014). The fiber volume fraction (V_f) of all composites was in the range of 0.25–0.27 having 3 layers of fabric with orientation of each layer in the same direction. In first category, composites were designated as U (untreated), A-0% (alkali-treated jute fabric with 0 wt% of PJF), 1, 5, and 10% (alkali-treated jute fabric with 1, 5, 10 wt% of PJF), respectively. In the second category, composite samples were designated as CF0 (uncoated), CF3 (3 wt% nanocellulose coated), CF5 (5 wt% nanocellulose coated), and CF10 (10 wt% nanocellulose coated), whereas third category composites were designated as untreated (untreated jute fabric), enzyme (enzyme-treated jute fabric), laser (laser-treated jute fabric), ozone (ozone-treated jute fabric), and plasma (plasma-treated jute fabric).

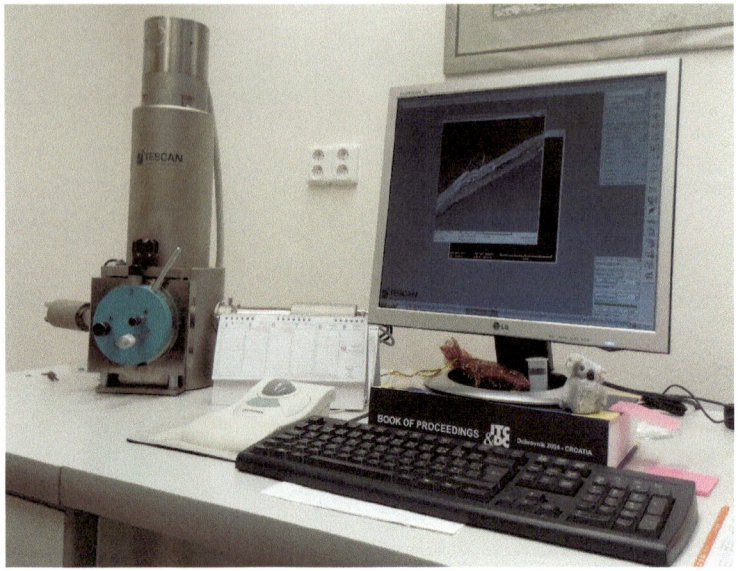

Fig. 3.7 Vega-Tescan TS5130 SEM

3.2.6 Characterization and Testing

3.2.6.1 Scanning Electron Microscopy (SEM)

The surface topologies of chemically treated and pulverized jute fibers, nanocellulose-coated jute fabrics, and surface-treated jute fabrics were observed with Vega-Tescan TS5130 scanning electron microscope (Fig. 3.7). The surface of fibers was gold coated prior to SEM inspection to improve the conductivity of samples. The pictures were taken at a slow scanning speed to obtain higher quality image.

3.2.6.2 Fourier Transform Infrared Spectroscopy (FTIR) and Particle Size Distribution

FTIR spectroscopy was done to confirm the removal of non-cellulosic contents (e.g., hemicellulose and lignin) from alkali-treated jute fabric and bleached jute fibers. A Thermo Fisher FTIR spectrometer, model Nicolet iN10 as shown in Fig. 3.8, was used in this study. The spectrometer was used in the absorption mode with a resolution of 4 cm^{-1}. Moreover, in order to measure the size (width) distribution of pulverized jute fibers and nanocellulose using SEM images, the topology of PJF and cellulose nanofibrils was also observed on Zeiss Ultra Plus field emission scanning electron microscope (FE-SEM) at low accelerating voltage

Fig. 3.8 Thermo
Fisher FTIR spectrometer

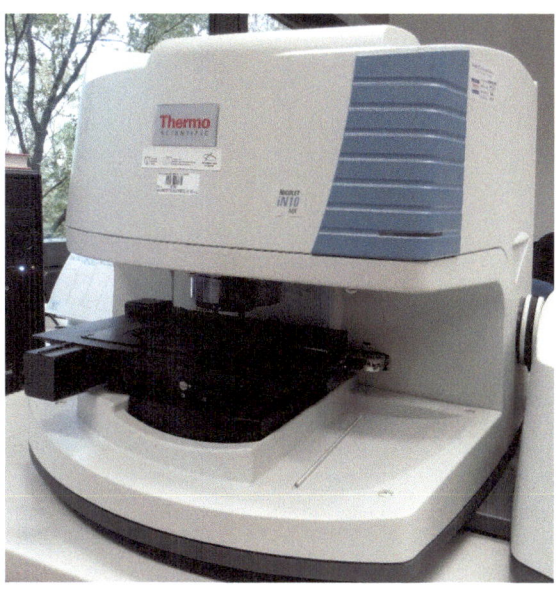

(1.0 kV) and low probe current (≈10 pA) to eliminate charging effect and sample damage due to interaction with primary electrons. The software used for image analysis was NIS Elements BR 3.22.

3.2.6.3 X-Ray Diffraction (XRD)

X-ray diffraction patterns were recorded on a PANalytical X' Pert PRO MPD diffraction system for untreated jute, bleached jute, and jute cellulose nanofibrils in order to examine the change in crystallinity of the material after bleaching and acid hydrolysis.

3.2.6.4 Mechanical Testing

Tensile and Flexural Tests

Tensile properties of first category of composites (incorporated with PJF) were measured on a universal testing machine, whereas second category of composites (reinforced with nanocellulose-coated jute fabrics) were characterized on an MTS series 370 servo-hydraulic load frame equipped with 647 hydraulic wedge grip of 100 kN load capacity (Fig. 3.9) at a cross head speed of 2 mm/min and gauge length of 100 mm in accordance with ASTM D3039-00 using rectangular specimens of dimension $200 \times 20 \times h$ mm^3, where "h" is the actual thickness of specimens. Flexural test was performed for all categories of composites in

Fig. 3.9 MTS series 370
servo-hydraulic loading
machine

three-point bending mode on a universal mechanical testing machine, Shimadzu
AGS-J with 5 kN load cell, following ASTM D790-03 standard at a cross head
speed of 2 mm/min (Fig. 3.11). The specimens of dimension $160 \times 12.7 \times h$ mm^3
were used maintaining a span-to-thickness ratio of 32:1 according to standard
recommendations ("h" is the actual thickness of specimen). Five specimens were
tested for each condition and for each test to get an average value.

Fatigue Test

The tension–tension fatigue performance of second category composites was
experimentally evaluated at two different stress levels (80 and 70% of the ultimate
tensile stress σ_u) assuming the same geometry of specimens as the one adopted in
quasi-static tensile tests. Tests were performed at gage length of 100 mm under
constant stress amplitude, stress ratio (R) of 0.1 (ratio of maximum to minimum
stress during a loading cycle), and frequency of 5 Hz. Three specimens were tested
for each stress level up to final failure of specimens. Tests were realized on the same
MTS series 370 servo-hydraulic loading machine equipped with 647 hydraulic
wedge grip of 100 kN load capacity as shown in Fig. 3.9.

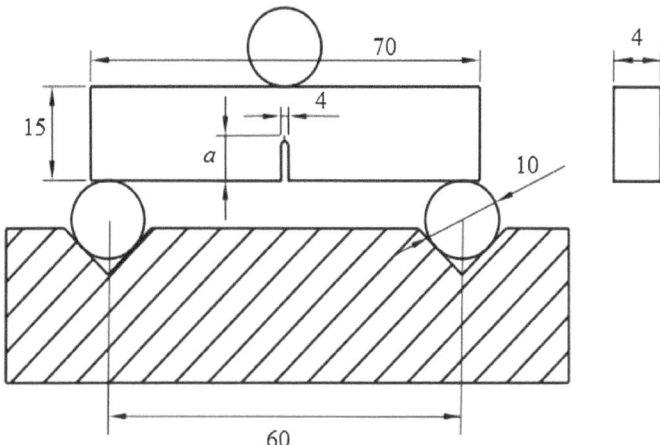

Fig. 3.10 SENB (single-edge-notch bending) specimen and fixture dimensions for fracture toughness test (dimensions in mm)

Fracture Toughness

Fracture toughness, K_{Ic}, of second category composites was determined by three-point bending method using the single-edge-notch bend (SENB) specimens in accordance with the standard test method ASTM D5045-99. A sharp crack of length "a" between 0.45 and 0.55 W was introduced by using a notch maker CEAST NOTCHVIS and a fresh razor blade at the notch tip ("W" is the width of specimen). The specifications of fixture and specimens are shown in Fig. 3.10.

The tests were performed using universal mechanical testing machine (Fig. 3.11), Shimadzu AGS-J with 5 kN load cell, at a cross head speed of 10 mm/min. Five specimens were tested for each condition to get an average value. Statistical analysis of tensile, flexural, and fracture toughness properties was done by one-way analysis of variance (ANOVA), and probability value $p \leq 0.05$ was considered as an indicative of statistical significance compared to the control samples.

3.2.6.5 Creep Test

Short-term creep tests were performed for only first category and third category of composites in three-point bending mode at temperatures 40, 70, and 100 °C for 30 min using Q800 dynamic mechanical analysis (DMA) instrument of TA instruments (New Castle DL, USA) as presented in Fig. 3.12. The static stress of 2.0 MPa was applied at the center point of long side of the sample through the sample thickness for 30 min after equilibrating at the desired temperature, and creep strain was measured as a function of time. The static stress was selected after

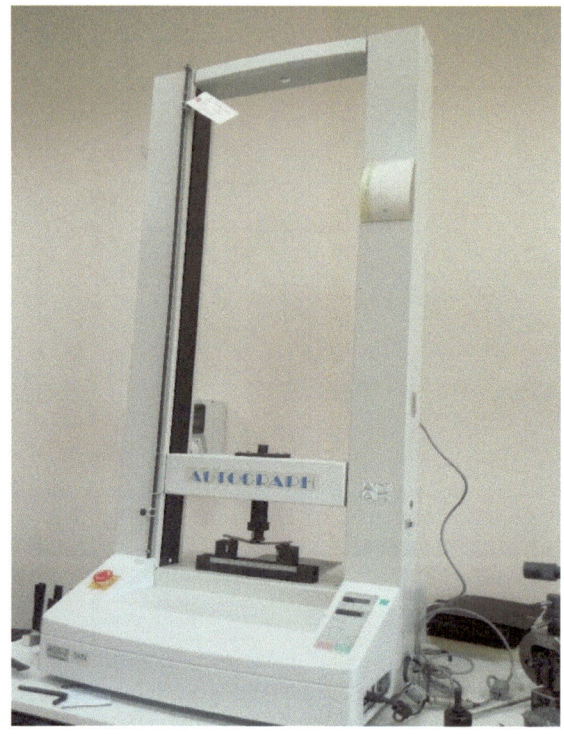

Fig. 3.11 Shimadzu AGS-J universal testing machine

Fig. 3.12 DMA Q800 instrument

performing a strain sweep test, where the linear viscoelastic region was defined for each of the composites ensuring that the creep tests were conducted in the linear viscoelastic region. The time–temperature superposition principle (TTSP) was selected for short-term creep tests performed at various temperatures for first category composites (jute/green epoxy composites incorporated with various contents of PJF). The temperature range was 40–100 °C, in 5 °C steps, and the isothermal tests were run on the same specimen in the specified temperature range. The 2.0 MPa stress was applied for 10 min at each temperature step. In every measurement, the specimen was equilibrated for 5 min at each temperature in order to evenly adjust for the correct temperature of the specimen.

3.2.6.6 Dynamic Mechanical Analysis

Dynamic mechanical analysis (DMA) is a technique to measure the mechanical properties of a material such as stiffness and damping as a function of temperature or time or both by applying a sinusoidal force at a set frequency. The stiffness is reported as modulus and the damping in reported as tan delta. Since the force is applied as oscillatory, the in-phase component is expressed as storage modulus (E'), and the out-phase component is expressed as loss modulus (E''). Storage modulus measures the elastic behavior of the material, and the viscous portion is measured in terms of loss modulus. Loss modulus represents the energy dissipated as heat. The ratio of loss to storage modulus is tan δ or damping (Fig. 3.13). It is a measure of how well a material can absorb energy. The dynamic mechanical properties of all composite categories were measured in three-point bending mode using the same instrument for creep testing. The testing conditions were controlled in the temperature range of 30–190 °C, with a heating rate of 3 °C/min, fixed frequency of 1 Hz, preload of 0.1 N, amplitude of 20 μm, and force track of 125%. The samples having thickness of 4–4.5 mm, width of 12 mm, and span length of 50 mm were used for both creep and DMA testing. Two replicate samples were tested for each test condition, and average values were reported.

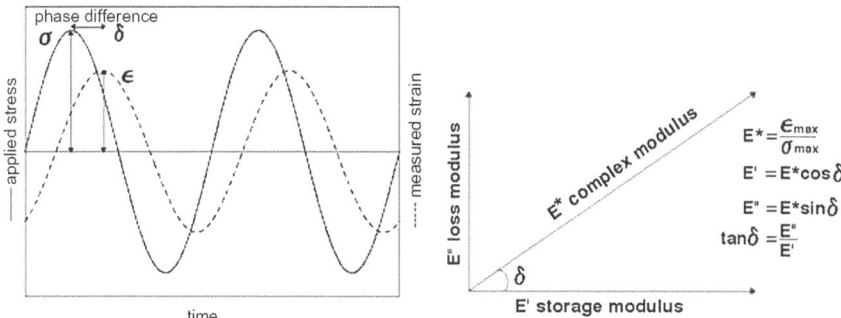

Fig. 3.13 The relationship of the applied sinusoidal stress to strain with the resultant phase lag and deformation

3.2.6.7 Creep Modeling

The nonlinear curve fit function of the OriginPro 9.0 software was used for modeling the creep curves and fitting the models to the experimental data. The minimum sum of squared deviation of experimental data from the creep models and coefficient of determination (R^2) were selected as criterion (Meloun and Militky 2011). The R^2 value is defined as model sum of squares divided by total sum of squares. A better goodness –of fit is obtained when R^2 is closer to 1. The Burger's model was used to simulate the short-term creep data of first category and third category of composites as well as long-term creep prediction of only first category of composites, whereas the Findley's power law model and two-parameter empirical power law model were used for long-term creep prediction of first category of composites.

References

Baheti, V., & Militky, J. (2013). Reinforcement of wet milled jute nano/micro particles in polyvinyl alcohol films. *Fibers and Polymers, 14*(1), 133–137.

Meloun, M., & Militky, J. (2011). *Statistical data analysis: A practical guide.* Sawston: Woodhead Publishing Limited.

Mishra, R., Baheti, V., Behera, B., & Militky, J. (2014). Novelties of 3-D woven composites and nanocomposites. *The Journal of the Textile Institute, 105*(1), 84–92.

Stepankova, M., Wiener, J., & Dembicky, J. (2010). Impact of laser thermal stress on cotton fabric. *Fibres & Textiles in Eastern Europe, 18*(3), 70–73.

Chapter 4
Effect of Pulverized Micro Jute Fillers Loading on the Mechanical, Creep, and Dynamic Mechanical Properties of Jute/Green Epoxy Composites

Abstract This chapter reports the mechanical, creep, and dynamic mechanical behavior of alkali-treated jute/green epoxy composites incorporated with various loadings (1, 5, and 10 wt%) of chemically treated pulverized jute fibers (PJF) at different environment temperatures. The tensile and flexural properties were improved with the incorporation of PJF in alkali-treated jute/green epoxy composites except the decrease in tensile strength of composite reinforced with only alkali-treated jute fabric. The creep and dynamic mechanical tests were performed in three-point bending mode by dynamic mechanical analyzer (DMA). The incorporation of PJF is also found to significantly improve the creep resistance and strain rate of composites. Three creep models, i.e., Burger's model, Findley's power law model, and a simpler two-parameter power law model, were used to model the creep behavior in this study. The time–temperature superposition principle (TTSP) was applied to predict the long-term creep performance. Findley's power law model was found to be satisfactory in predicting the long-term creep behavior. Dynamic mechanical thermal analysis (DMTA) results revealed the increase in storage modulus, glass transition temperature, and reduction in the tangent delta peak height of composites with higher loading of PJF.

Keywords Pulverized jute fibers (PJF) · Green epoxy · Creep · Time temperature superposition principle (TTSP)

4.1 Overview

In this chapter, the characterization of alkali-treated jute fabric and chemically pretreated waste jute fibers by FTIR is presented. The particle size distribution and surface topology of pulverized micro jute fillers by SEM are shown. The tensile, flexural creep and dynamic mechanical properties of PJF-filled jute/green epoxy composites are measured. Three creep models i.e., Burger's model, Findley's power law model, and a simple two-parameter power law model are used to model the

© The Author(s) 2017   57
A. Jabbar, *Sustainable Jute-Based Composite Materials*, SpringerBriefs in Applied
Sciences and Technology, DOI 10.1007/978-3-319-65457-7_4

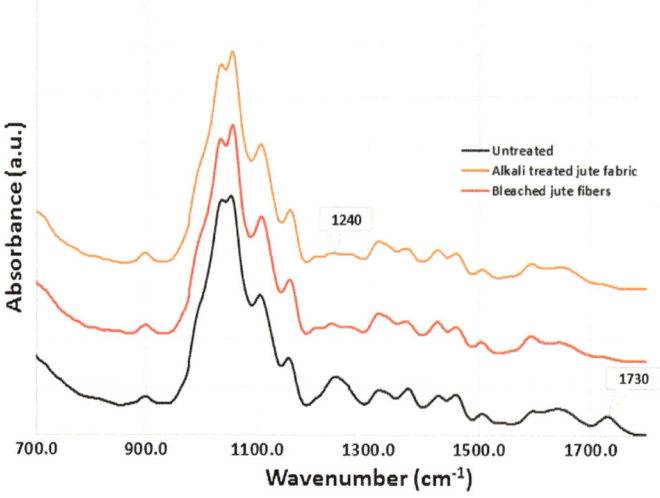

Fig. 4.1 FTIR spectra of jute fibers

creep behavior of composites. The time–temperature superposition principle
(TTSP) is applied to predict the long-term creep performance of composites.

4.2   Results and Discussion

4.2.1   Characterization of Jute Fibers

FTIR analysis was carried out to confirm some removal of non-cellulosic contents
(e.g., hemicelluloses and lignin) from the surface of jute fibers after pretreatments.
The FTIR spectra of untreated and treated jute are shown in Fig. 4.1. The major
difference observed between the spectra is the disappearance/reduction of peaks at
~ 1730 cm^{-1} and ~ 1240 cm^{-1}. The peak at ~ 1730 cm^{-1} is due to stretch
vibration of C=O bonds in carboxylic acid and ester components of cellulose and
hemicellulose (Morshed et al. 2010) and also carbonyl group of lignin (Tserki et al.
2005; Haque et al. 2009). The peak at ~ 1240 cm^{-1} is due to C–O–C asymmetric
stretching of the acetyl group of lignin (Liu et al. 2004). The reduction/
disappearance of these peaks confirms the partial removal of hemicellulose and
opening up of the lignin structure in the jute fibers after pretreatments.

The SEM image is precisely analyzed to measure the size (diameter) of pulverized jute fibers (PJF), after 1 h of milling, as shown in Fig. 4.2a. The histogram
of size distribution of 100 measurements is shown in Fig. 4.2b. The calculated
average diameter (width) of PJF was found to be 1.856 ± 0.899 μm.

(a)

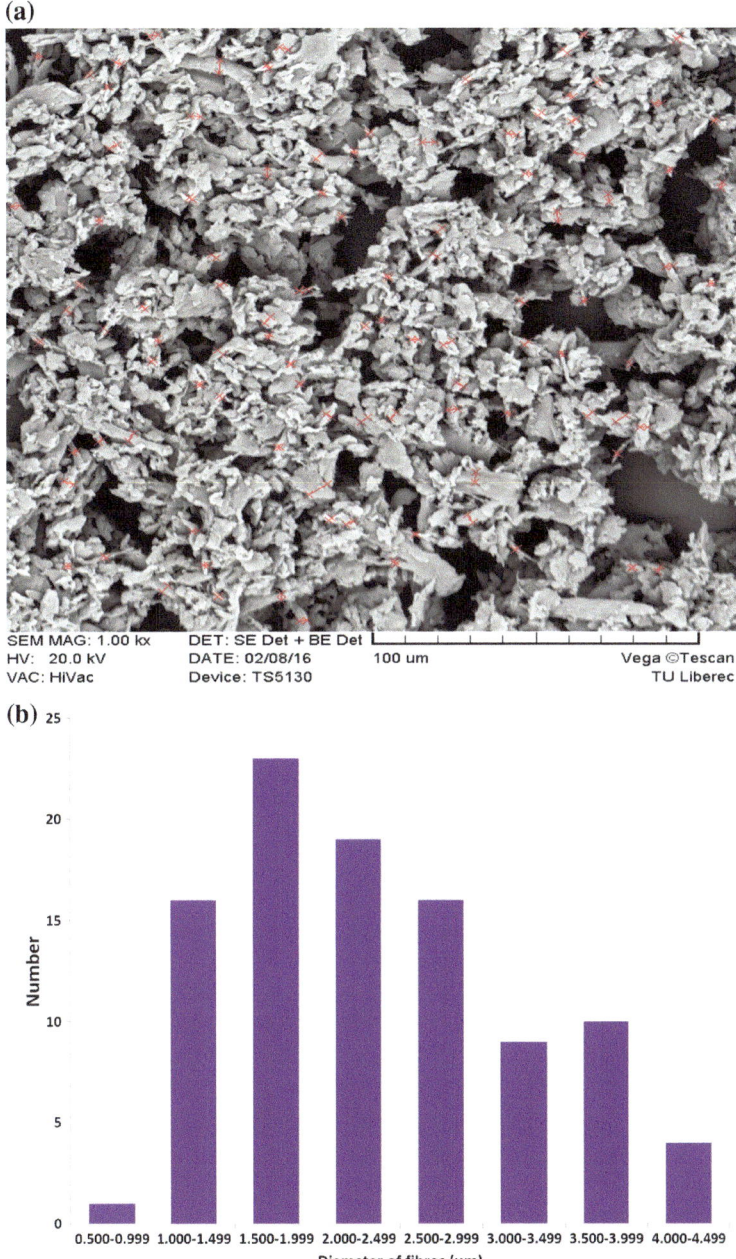

(b)

Fig. 4.2 **a** SEM image of jute fibers after 1.0 h of pulverization and **b** histogram of particle width distribution

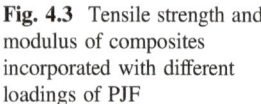

Fig. 4.3 Tensile strength and modulus of composites incorporated with different loadings of PJF

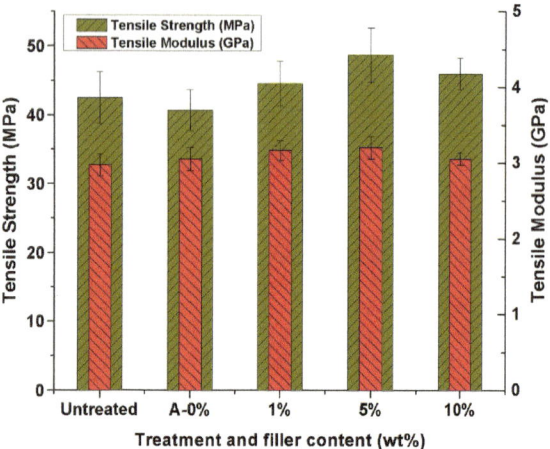

4.2.2   Tensile Properties

Figure 4.3 shows the average values and standard deviation of ultimate tensile strength and tensile modulus of alkali-treated jute composites with different loadings of PJF. It is interesting to note that alkali treatment of jute fabric has resulted in a lowering of the tensile strength, but increase in tensile strength and tensile modulus is observed with the incorporation of PJF. The tensile strength decreases from 42.48 MPa for untreated composite to 40.65 MPa for alkali-treated composite, thus presenting 4.3% decrease on average. However, tensile strength improves by factors of 4.9, 14 and 8% with 1, 5 and 10% loading of PJF, respectively, as compared to untreated composite. Similarly, tensile modulus increases from 2.97 GPa for untreated composite to 3.05, 3.16, 3.20, and 3.05 GPa for A-0%, 1, 5, and 10% composites, respectively, thus permitting 2.7, 6.3, 7.7, and 2.7% increase on average as shown in Fig. 4.3.

We must keep in mind that the mechanics of textile composites is different from those of short fiber composites. The major contribution to strength in textile composites is the alignment of yarns in warp and weft direction. Alkali treatment results in the partial unwinding of yarns (as hemicellulose dissolves), and hence, the alignment gets antagonized. This results in a lowering of the strength of composites (Jacob et al. 2006). The incorporation of PJF as filler in matrix provides a better reinforcing effect, thus improving fiber/matrix interfacial interactions in composites and hence tensile properties.

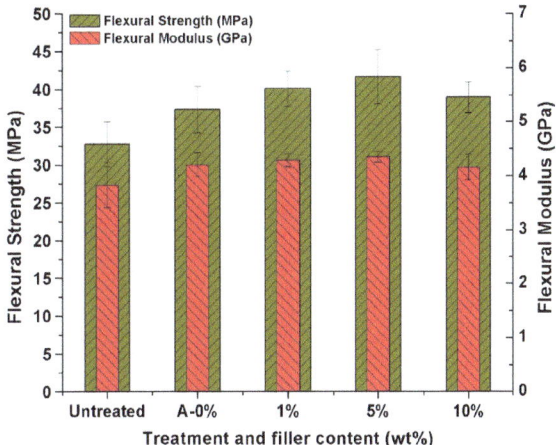

Fig. 4.4 Flexural strength and modulus of composites incorporated with different loadings of PJF

4.2.3 Flexural Properties

The flexural strength and modulus increase with the increase in contents of PJF in alkali-treated jute composites as shown in Fig. 4.4. The flexural strength and modulus increase from 32.78 MPa and 3.83 GPa for untreated composite to maximum of 41.66 MPa and 4.35 GPa for composite loaded with 5 wt% of PJF, respectively, thus allowing 27 and 13.6% increase on average as shown in Fig. 4.4. These findings also suggest better reinforcing effect provided by PJF, thus contributing to strong interfacial interaction of fiber and matrix. However, the reason for a little reduction in the tensile and flexural properties of composite for 10 wt% loading of PJF (compared to 5% composite) may be due to the aggregation of PJF, which generates defects in the material. Stress concentration is likely to occur within the resin or agglomerated particles, which could generate slippage within the material because of the external force, resulting in reduction tensile properties (Kargarzadeh et al. 2015).

4.2.4 Short-Term Creep

Figure 4.5 shows the creep strains for jute composites as a function of time with 0, 1, 5, and 10 wt% of PJF content at three different temperature conditions. It is visibly apparent that the composites have low instantaneous deformation ε_M and creep strain at 40 °C due to higher stiffness of composites, but this deformation increases at higher temperatures due to decrease in composites stiffness. The creep strain of all composites also increased at higher temperatures, but the untreated jute composites was affected more than the others. The creep strain of alkali-treated with 0% PJF composite is less than untreated one. This may be explained due to increase

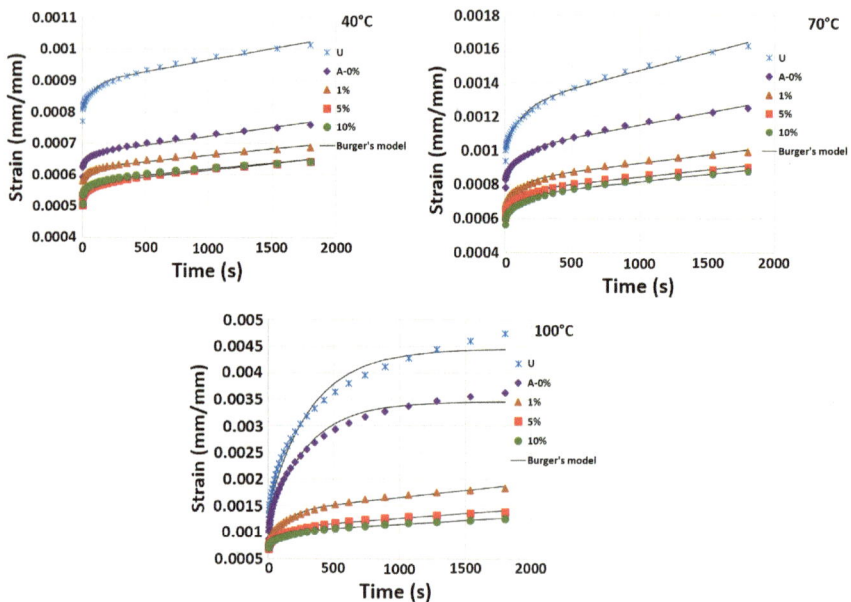

Fig. 4.5 Creep curves of composites incorporated with different loadings of PJF at different temperatures

in surface roughness of jute fabric after alkali treatment and decrease in frictional slippage of matrix polymer chains at the fiber/matrix interface resulting in less creep deformation than untreated composite. The least creep strain is shown by composite incorporated with 10% PJF at all temperatures followed by 5 and 1% PJF incorporated composites. At 100 °C, 5 and 10% PJF composites have almost same instantaneous elastic deformation, but 10% composite has less viscous deformation over time. This may be attributed to greater inhibition of slippage and reorientation of polymer chain with increasing contents of PJF. The Burger's model curves show a satisfactory agreement with the experimental data (Fig. 4.5).

The four parameters E_M, E_K, η_M, η_K of Burger's model, used to fit Eq. 2.3 to the experimental data, are summarized in Table 4.1. The first value is parameter estimator, and value in parenthesis is corresponding standard deviation. All four parameters were found to decrease for all composites as temperature increased (Table 4.1). E_M corresponds to the elasticity of the crystallized zones in a semicrystallized polymer. Compared to the amorphous regions, the crystallized zones are subjected to immediate stress due to their higher stiffness. The instantaneous elastic modulus is recovered immediately once the stress is removed. E_K is also coupled with the stiffness of material. The decrease in parameters E_M and E_K resulted from the increase in the instantaneous and the viscoelastic deformations as temperature increased. The viscosity η_M corresponds to damage in the crystallized zones and irreversible deformation in the amorphous regions, and the viscosity η_K is also associated with the viscosity of the amorphous regions in the semicrystallized

Table 4.1 Simulated four parameters in Burger's model for short-term creep of the composites

Temperature	Parameters	Composite types				
		Untreated	Alkali-0%	1%	5%	10%
40 °C	E_m [MPa]	2477.24 (78.3)	3259.41 (138.0)	3492.53 (133.8)	3810.53 (133.6)	3774.75 (149.1)
	E_k [MPa]	23876.27 (10854.5)	38244.62 (19726.9)	42496.38 (21110.6)	44220.12 (21281.1)	40549.80 (19016.6)
	η_m [Pa.s]	2.72E7 (1.33E7)	3.54E7 (1.22E7)	4.77E7 (2.04E7)	4.54E7 (1.97E7)	5.11E7 (2.47E7)
	η_k [Pa.s]	2.11E6 (2.38E6)	1.37E6 (2.09E6)	1.73E6 (2.5E6)	2.53E6 (3.49E6)	1.90E6 (2.62E6)
	SS^*	2.65033E-9	1.29037E-9	1.04302E-9	9.92292E-10	1.08146E-9
	$Adj. R^2$	0.97524	0.97061	0.96424	0.96929	0.96304
70 °C	E_m [MPa]	1985.89 (91.9)	2403.61 (102.7)	2921.19 (118.5)	3116.67 (118.6)	3323.10 (131.9)
	E_k [MPa]	7790.84 (2752.5)	11972.90 (4497.3)	14228.47 (5360.3)	16946.57 (6037.4)	15615.45 (5726.5)
	η_m [Pa.s]	9.48E6 (3.81E6)	1.32E7 (5.14E6)	1.98E7 (9.71E6)	2.33E7 (1.08E7)	2.27E7 (1.13E7)
	η_k [Pa.s]	956179.43 (6.90E5)	1.32E6 (1.09E6)	1.71E6 (1.34E6)	1.81E6 (1.45E6)	1.94E6 (1.44E6)
	SS^*	1.08937E-8	5.93037E-9	3.81543E-9	2.74811E-9	2.87718E-9
	$Adj. R^2$	0.98881	0.98689	0.9851	0.9847	0.98596
100 °C	E_m [MPa]	1380.05 (294.3)	1743.43 (306.1)	2417.83 (218.3)	2653.56 (183.6)	2616.45 (123.7)
	E_k [MPa]	665.35 (117.7)	865.63 (121.6)	3543.54 (958.0)	6087.72 (1783.7)	8457.41 (2683.5)
	η_m [Pa.s]	−4.13E20 (0.0)	−6.00E32 (0.0)	7.44E6 (3.92E6)	1.08E7 (5.26E6)	1.31E7 (5.97E6)
	η_k [Pa.s]	219089.75 (1.03E5)	255733.51 (9.91E4)	457813.67 (2.43E5)	705184.01 (4.39E5)	1.11E6 (6.84E5)
	SS^*	8.30033E-7	3.37586E-7	2.87859E-8	1.31777E-8	6.79893E-9
	$Adj. R^2$	0.97565	0.98358	0.99023	0.98826	0.98971

SS^* Sum of squared deviations

polymer (Militký and Jabbar 2015). The decrease in viscosity parameters η_M, η_K proposes an improvement in the mobility of molecular chains at higher temperature. The parameters for untreated and alkali-treated with 0% PJF composites have undergone a largest decrease, resulting in higher creep strain. The composites incorporated with PJF, especially 5 and 10%, have comparatively better values of parameters particularly η_M, which is related to the long-term creep strain and validates less temperature dependence of these composites (Fig. 4.5). The viscosity η_M increases with the increase in PJF % and permanent deformation decreases. Figure 4.6a, b compares the creep strain and strain rate of untreated and 10% PJF composites at various temperatures. Comparatively, temperature had more influence on the creep deformation of untreated jute composite than that of 10% PJF composite.

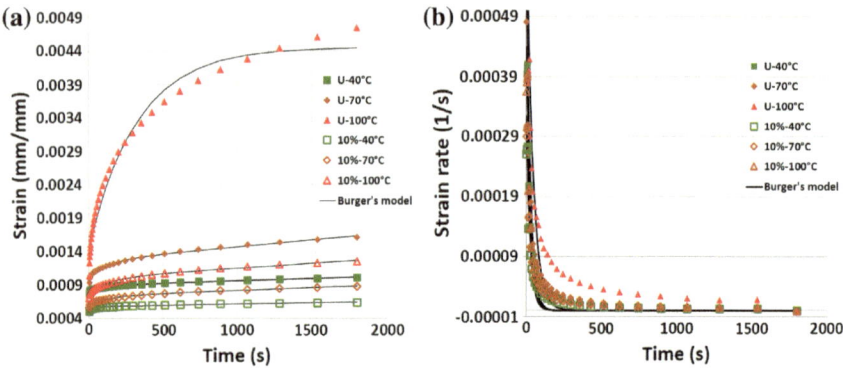

Fig. 4.6 Creep strain (**a**) and strain rate (**b**) of untreated and 10% PJF composites at different temperatures

4.2.5 Time–Temperature Superposition (TTS)

The creep curves corresponding to different temperature levels were shifted along the logarithmic time axis according to time–temperature superposition principle using TA Instruments Thermal Advantage™ software to generate a master curve at a reference temperature of 40 °C. The shifting procedure of this curve obeys the Williams–Landel–Ferry (WLF) equation. The WLF equation is given by Eq. 4.1;

$$\log \alpha_T = \frac{-C_1(T - T_0)}{C_2 + (T - T_0)} \tag{4.1}$$

where α_T is the horizontal (or time) shift factor, C_1 and C_2 are constants, T_0 is the reference temperature (K), and T is the test temperature (K).

The master curves, which give an indication of long-term creep performance of composites, are plotted in log–log scale and presented in Fig. 4.7. The master curves show better creep resistance of composites with increasing content of PJF. It is obvious that the best long-term performance is shown by composites incorporated with 5 and 10% PJF indicating their good reinforcing effectiveness. It is also interesting to note that above log-time 4.0 s, the creep deformation of untreated 0 and 1% composites show a faster tendency of increase compared to 5 and 10% PJF composites. These findings show that under the small stress, the materials entered into a viscoelastic state over an extremely long period of time, and in viscoelastic state, the role of 1% incorporation of PJF in the reinforcement effectiveness is less than that of 5 and 10% PJF.

The simulated parameters of Burger, Findley, and two-parameter power law models are summarized in Table 4.2. The first value is parameter estimator, and value in parenthesis is corresponding standard deviation. Based on the sum of squared deviations (SS) and R^2 values, it can be clearly seen that Findley's model is good to predict the long-term creep performance as compared to Burger's and

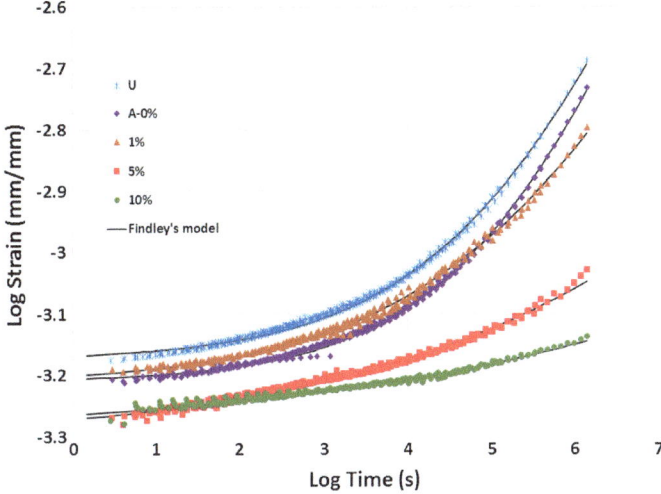

Fig. 4.7 TTS master curves for creep of the composites incorporated with different loadings of PJF at a reference temperature of 40 °C

two-parameter power law model. It is also shown in Fig. 4.7 that the prediction ability of the Findley's model is good for the long-term creep behavior of composites.

However, this model provides the adequate prediction ability within the steady-state creep and given time interval as revealed by some researchers (Siengchin and Karger-Kocsis 2009; Siengchin 2013) while for longer time duration, the calculated data may show considerable deviation from the experimental data. The sum of squared deviations and R^2 values, given in Table 4.2, also suggest that the Burger's model shows some deviation and that of two-parameter power law model shows a little large deviation from the experimental data. Similar findings were reported by other researchers (Jia et al. 2011; Hao et al. 2014).

The parameters of Burger's model, resulted from fitting master creep curves, are very different from those of the short-term creep tests. It can be seen from Table 4.2 that all the Burger's model parameters increase with the increase in loading of PJF %. The η_M, which determines long-term creep, is the lowest for untreated composite and highest for 10% PJF composite. Therefore, the untreated composite shows the highest, and 10% PJF composite shows the lowest creep deformation. It is also obvious for Findley model that the parameter a (reflecting short-term creep) increased and parameters ε_0 (reflecting the instantaneous initial creep strain) and b (reflecting log-term creep) decreased with the increasing content of PJF which indicates an enhanced long-term creep performance with PJF loading. Similarly, for two-parameter power law model, parameter a (reflecting short-term creep) increased and parameter b (reflecting log-term creep) decreased with the increasing content of PJF.

Table 4.2 Simulated parameters of Burger's model, Findley's power law model, and two-parameter power law model for long-term creep prediction of the composites at 40 °C

Temperature	Parameters	Composite types				
		Untreated	Alkali-0%	1%	5%	10%
Burger's model	E_m [MPa]	2713.76 (193.6)	3002.45 (184.0)	2904.39 (179.0)	3478.01 (191.0)	3517.07 (125.5)
	E_k [MPa]	4781.98 (2005.8)	5470.29 (2194.5)	6271.53 (2337.0)	15949.06 (7113.4)	31490.67 (16174.2)
	η_m [Pa.s]	2.64E9 (1.29E9)	2.94E9 (1.29E9)	4.01E9 (2.19E9)	9.44E9 (6.93E9)	2.10E10 (2.04E10)
	η_k [Pa.s]	1.0227E8 (8.73E7)	1.35E8 (1.05E8)	9.24E7 (7.83E7)	6.76E7 (9.00E7)	9.49E7 (1.53E8)
	SS^*	4.86566E-7	3.00571E-7	2.96188E-7	1.2249E-7	4.56987E-8
	Adj. R^2	0.9696	0.97551	0.9658	0.92279	0.88442
Findley's model	a	1.05E-5 (4.83E-06)	6.49E-6 (3.69E-06)	1.79E-5 (1.14E-05)	3.49E-5 (3.45E-05)	3.35E-5 (5.92E-04)
	b	0.34341 (3.39E-02)	0.3699 (4.22E-02)	0.2803 (4.63E-02)	0.17143 (6.61E-02)	0.1284 (2.27E00)
	ε_0	6.70E-4 (2.68E-05)	6.17E-4 (2.58E-05)	6.13E-4 (3.82E-05)	5.00E-4 (5.74E-05)	5.11E-4 (9.03E-03)
	SS^*	3.83097E-8	4.27255E-8	4.37949E-8	1.91548E-8	1.30874E-8
	Adj. R^2	0.99761	0.99653	0.99496	0.98797	0.96701
Two-parameter power law model	a	4.63E-4 (1.41E-04)	4.24E-4 (1.39E-04)	4.82E-4 (1.00E-04)	4.85E-4 (3.61E-05)	5.22E-4 (2.08E-05)
	b	0.08443 (3.37E-02)	0.08145 (3.64E-02)	0.06828 (2.37E-02)	0.03723 (9.03E-03)	0.01951 (5.01E-03)
	SS^*	2.49937E-6	2.31209E-6	1.0253E-6	9.22041E-8	2.50976E-8
	Adj. R^2	0.8452	0.8128	0.88238	0.94226	0.93694

SS^* Sum of squared deviations

4.2.6   Dynamic Mechanical Properties

Dynamic mechanical analysis can characterize the viscoelastic properties of the materials and determine the information of storage modulus, loss modulus (the energy dissipation associated with the motion of polymer chains), and loss factor (tan delta) of polymer composites within the measured temperature range (Wang et al. 2015). The variation of storage modulus (E') of composites incorporated with different content of PJF as a function of temperature at frequency of 1 Hz is shown in Fig. 4.8.

It can be seen from Fig. 4.8a that there is a gradual fall in the storage moduli with temperature, which should be related with an energy dissipation phenomenon involving cooperative motions of the polymer chains with temperature (Feng et al. 2011). The increase in storage modulus over the whole temperature range was observed for composites incorporated with different loadings of PJF, for example, addition of 1, 5, and 10% PJF causes a significant increase of $\sim$18, 22 and 43% in the storage modulus, respectively, at 35 °C. Moreover, the storage modulus curves of composites have been shifted to higher temperatures after addition of the PJF, particularly 5 and 10% loading. This significant improvement in storage modulus is

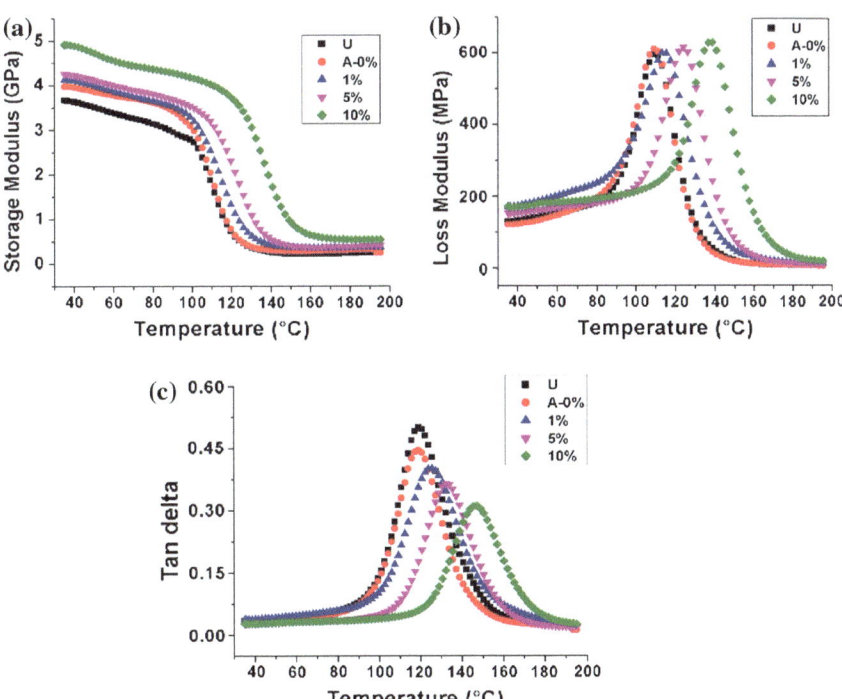

Fig. 4.8 Dynamic mechanical properties of composites incorporated with different loadings of PJF; **a** storage modulus, **b** loss modulus, and **c** tan delta

Table 4.3 T_g values
obtained from E'' curves

Composites	T_g from E''_{max} curve [°C]
Untreated	110.10
Alkali-0%	110.60
1%	114
5%	123
10%	137

due to better reinforcing effect of PJF leading to increased stiffness and the mobility restriction of polymer chains (Jia et al. 2011).

The change in loss factor (tan δ, the ratio of loss modulus to corresponding storage modulus) of composites with different loading of PJF as a function of temperature is shown in Fig. 4.8c. Untreated composite displayed a higher tanδ peak value than others. This may be attributed to more energy dissipation due to frictional damping at the weaker fiber/matrix interface. The temperature at which tanδ attains a maximum value can be referred to as the glass transition temperature (T_g) (Shanmugam and Thiruchitrambalam 2013). A positive shift in T_g can be observed for all composites incorporated with PJF compared to untreated composite. The lower tan δ peak height is shown by composite incorporated with 10% PJF followed by 5 and 1% PJF composites, exhibiting a strong fiber/matrix interfacial interactions which can restrict the segmental movement of the polymer chains leading to the increased T_g.

It has been reported that T_g values obtained from loss modulus (E'') curve peak are more realistic as compared to those obtained from loss factor (tanδ) (Akay 1993). A positive shift in T_g to higher temperature for all composites incorporated with PJF is observed, i.e., T_g increased from 110.1 °C for untreated to $\sim$110.6, 114, 123 and 137 °C for composites incorporated with 0, 1, 5, and 10% PJF, respectively, as presented in Table 4.3 and Fig. 4.8b. This may be due to reduced mobility of matrix polymer chains and better reinforcement effect of PJF. It has been reported that systems containing more restrictions and a higher degree of reinforcement tend to exhibit higher T_g (Almeida et al. 2012).

4.3 Summary

The tensile and flexural properties were found to improve with the incorporation of PJF in alkali-treated jute/green epoxy composites except the decrease in tensile strength of composite reinforced with only alkali-treated jute fabric. The creep deformation was seen to increase with temperature. The creep resistance of composites was found to improve significantly with the incorporation of PJF. The modeling of creep data was satisfactorily conducted by using Burger's model, Findley's power law model, and a simpler two-parameter power law model. The Burger's model fitted well the short-term creep data. However, the Findley's power law model was satisfactory for predicting the long-term creep performance of composites compared to Burger and two-parameter models. The master curves, generated by TTSP, indicated the prediction of the long-term performance of composites. Dynamic mechanical test results revealed the increase in storage modulus, glass transition temperature, and reduction in the tangent delta peak height of PJF incorporated composites. Based on the analysis of results, the improved creep resistance of the composites was likely attributed to the inhibited mobility of polymer matrix molecular chains initiated by large interfacial contact area of PJF as well as their interfacial interaction with the polymer matrix.

References

Akay, M. (1993). Aspects of dynamic mechanical analysis in polymeric composites. *Composites Science and Technology, 47*(4), 419–423.

Almeida, J. H. S., Jr., Ornaghi, H. L., Jr., Amico, S. C., & Amado, F. D. R. (2012). Study of hybrid intralaminate curaua/glass composites. *Materials and Design, 42,* 111–117.

Feng, Q.-P., Shen, X.-J., Yang, J.-P., Fu, S.-Y., Mai, Y.-W., & Friedrich, K. (2011). Synthesis of epoxy composites with high carbon nanotube loading and effects of tubular and wavy morphology on composite strength and modulus. *Polymer, 52*(26), 6037–6045.

Hao, A., Chen, Y., & Chen, J. Y. (2014). Creep and recovery behavior of kenaf/polypropylene nonwoven composites, *Journal of Applied Polymer Science, 131*(17). doi:10.1002/app.40726.

Haque, M. M., Hasan, M., Islam, M. S., & Ali, M. E. (2009). Physico-mechanical properties of chemically treated palm and coir fiber reinforced polypropylene composites. *Bioresource Technology, 100*(20), 4903–4906.

Jacob, M., Thomas, S., & Varughese, K. (2006). Novel woven sisal fabric reinforced natural rubber composites: Tensile and swelling characteristics. *Journal of Composite Materials, 40* (16), 1471–1485.

Jia, Y., Peng, K., Gong, X.-L., & Zhang, Z. (2011). Creep and recovery of polypropylene/carbon nanotube composites. *International Journal of Plasticity, 27*(8), 1239–1251.

Kargarzadeh, H., Sheltami, R. M., Ahmad, I., Abdullah, I., & Dufresne, A. (2015). Cellulose nanocrystal: A promising toughening agent for unsaturated polyester nanocomposite. *Polymer, 56,* 346–357.

Liu, W., Mohanty, A., Drzal, L., Askel, P., & Misra, M. (2004). Effects of alkali treatment on the structure, morphology and thermal properties of native grass fibers as reinforcements for polymer matrix composites. *Journal of Materials Science, 39*(3), 1051–1054.

Militký, J., & Jabbar, A. (2015). Comparative evaluation of fiber treatments on the creep behavior of jute/green epoxy composites. *Composites Part B Engineering, 80,* 361–368.

Morshed, M., Alam, M., & Daniels, S. (2010). Plasma treatment of natural jute fibre by RIE 80 plus plasma tool. *Plasma Science and Technology, 12*(3), 325.

Shanmugam, D., & Thiruchitrambalam, M. (2013). Static and dynamic mechanical properties of alkali treated unidirectional continuous palmyra palm leaf stalk fiber/jute fiber reinforced hybrid polyester composites. *Materials and Design, 50,* 533–542.

Siengchin, S. (2013). Dynamic mechanic and creep behaviors of polyoxymethylene/boehmite alumina nanocomposites produced by water-mediated compounding effect of particle size. *Journal of Thermoplastic Composite Materials, 26*(7), 863–877.

Siengchin, S., & Karger-Kocsis, J. (2009). Structure and creep response of toughened and nanoreinforced polyamides produced via the latex route: Effect of nanofiller type. *Composites Science and Technology, 69*(5), 677–683.

Tserki, V., Zafeiropoulos, N., Simon, F., & Panayiotou, C. (2005). A study of the effect of acetylation and propionylation surface treatments on natural fibres. *Composites Part A Applied Science and Manufacturing, 36*(8), 1110–1118.

Wang, X., Gong, L.-X., Tang, L.-C., Peng, K., Pei, Y.-B., Zhao, L., et al. (2015). Temperature dependence of creep and recovery behaviors of polymer composites filled with chemically reduced graphene oxide. *Composites Part A Applied Science and Manufacturing, 69,* 288–298.

Chapter 5
Extraction of Nanocellulose from Waste Jute Fibers and Characterization of Mechanical and Dynamic Mechanical Behavior of Nanocellulose-Coated Jute/Green Epoxy Composites

Abstract The work presented in this chapter was aimed to explore the effect of nanocellulose coating on the mechanical and thermomechanical properties of jute/green epoxy composites. Cellulose was purified from waste jute fibers and converted to nanocellulose by acid hydrolysis, and subsequently, 3, 5, and 10 wt% of nanocellulose suspensions were coated over woven jute reinforcement to prepare composites. The surface topologies of treated jute fibers, jute cellulose nanofibrils (CNF), nanocellulose-coated jute fabrics, and fractured surfaces of composites were characterized by SEM. Composites were evaluated for tensile, flexural, fatigue, fracture toughness, and dynamic mechanical properties. The results revealed the improvement in composite properties such as tensile modulus, flexural strength, flexural modulus, fatigue life, and fracture toughness with the increase in the concentration of nanocellulose coating over jute reinforcement except the decrease in tensile strength. The storage modulus was increased, and tangent delta peaks were reduced for nanocellulose-coated jute composites as presented by DMA results.

Keywords Nanocellulose coating · Acid hydrolysis · Mechanical properties · Fracture toughness · Thermo-mechanical analysis

5.1 Overview

In this chapter, the preparation and characterization of nanocellulose-coated woven jute/green epoxy composites are presented. Waste jute fibers are used as precursor to purify and extract nanocellulose by chemical treatments. The prepared suspensions with 3, 5, and 10 wt% of nanocellulose are coated over jute fabric followed by preparation of composites by hand layup and compression molding technique. The surface topologies of chemically treated jute fibers, jute cellulose nanofibrils, nanocellulose-coated jute fabrics, and fractured surfaces of composites are characterized by scanning electron microscopy (SEM). The crystallinity of jute fibers after chemical treatments is measured by X-ray diffraction (XRD). The effect of

© The Author(s) 2017   71
A. Jabbar, *Sustainable Jute-Based Composite Materials*, SpringerBriefs in Applied
Sciences and Technology, DOI 10.1007/978-3-319-65457-7_5

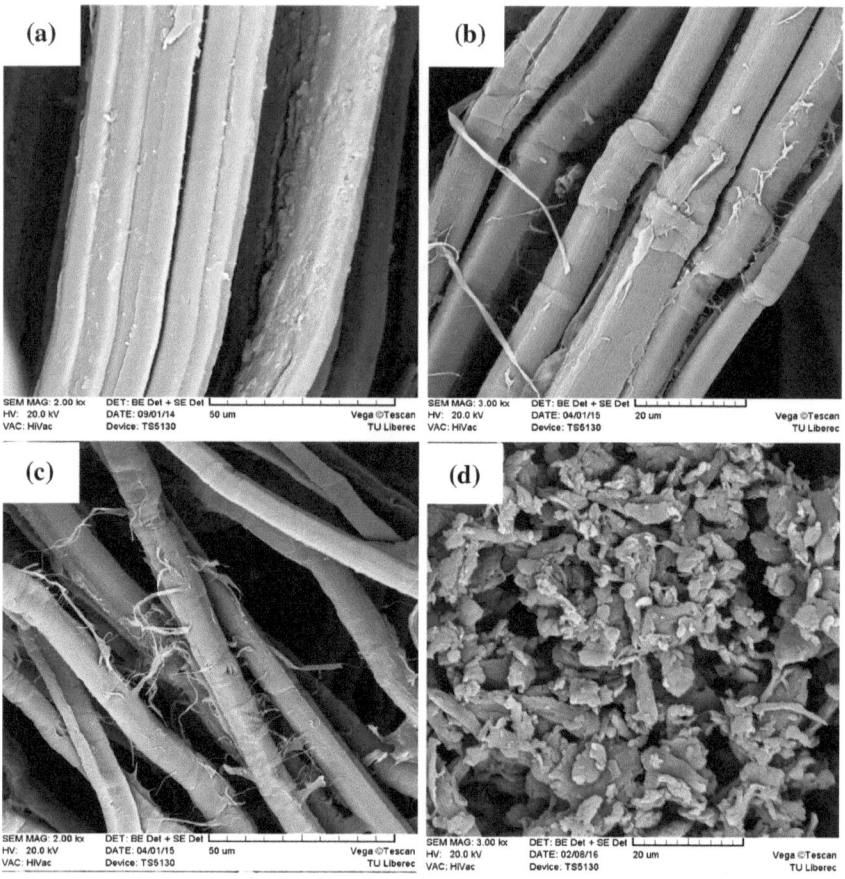

Fig. 5.1 SEM images of jute fibers **a** untreated, **b** alkali treated, **c** bleached, and **d** pulverized (milled)

nanocellulose coating over woven jute reinforcement on the tensile, flexural, fatigue, fracture toughness, and dynamic mechanical properties of composites has been investigated.

5.2 Results and Discussion

5.2.1 SEM Study of Chemically Treated Jute Fibers and Jute Cellulose Nanofibrils

Surface topologies of jute fibers after alkali treatment, bleaching, and milling are examined by SEM and presented in Fig. 5.1. Figure 5.1a shows the jute fibers bound together in the form of fiber bundles by cementing materials, e.g., hemicellulose and lignin, but after repeated alkali treatment, splitting of fibers is observed due to destruction of mesh structure with a little rough and clean surface, maybe due to majority of the removal of hemicellulose, lignin, and other non-cellulosic materials (Mukherjee et al. 1993; Ray et al. 2001) as shown in Fig. 5.1b.

The alkali-treated fibers are further separated into individual fibers with more clean surface and fibrillation on the surface, maybe due to more delignification after

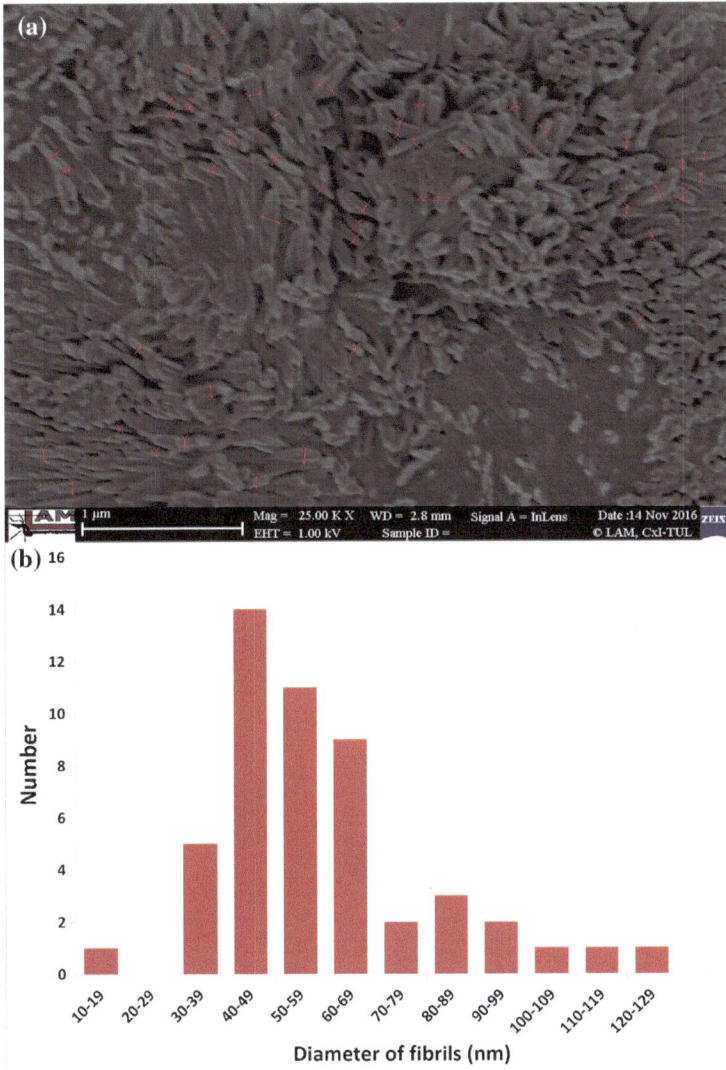

Fig. 5.2 a FE-SEM image of jute cellulose nanofibrils. **b** Histogram of width distribution of cellulose nanofibrils

bleaching treatment (Beg and Pickering 2008), as shown in Fig. 5.1c. Figure 5.1d displays the milled jute fibers with size distribution of fibers in the micron range.

A high-resolution FE-SEM image at nanoscale level is precisely analyzed to measure the size (width) of jute cellulose nanofibrils (CNF), after acid hydrolysis, as shown in Fig. 5.2a. The histogram of size distribution of 50 measurements is shown in Fig. 5.2b. The calculated average diameter (width) of CNF was found to be 57.40 ± 20.61 nm.

5.2.2 Surface Topology of Nanocellulose-Coated Jute Fabric

The topological changes that occur on the surface of jute fabric after nanocellulose coating are shown in Fig. 5.3. It is apparent that there is depositing of nanocellulose on the fabric surface forming a layer. The layer thickness increases gradually with the increase in nanocellulose concentration as clear in Fig. 5.3b–d, and the fabric surface coated with 10 wt% nanocellulose suspension is covered almost completely with cellulose as shown in Fig. 5.3d.

5.2.3 XRD Analysis of Jute Fibers

The X-ray diffraction patterns of untreated, bleached jute fibers, and jute CNF are shown in Fig. 5.4. These diffraction patterns are typical of semicrystalline materials with an amorphous broad small hump and a large crystalline peak. Two well-defined peaks around $2\theta \approx 16°$ and $23°$ are typical of cellulose-I. The crystallinity index for all samples was determined by using the Eq. 5.1 (Terinte et al. 2011):

$$\text{Crystallinity } \% = 100 \times \frac{I_{200} - I_{\text{non-cr}}}{I_{200}}, \tag{5.1}$$

where I_{200} represents maximum intensity of the peak corresponding to the plane 200 at 2θ angle between 22 and 24° and $I_{\text{non-cr}}$ is the intensity for diffraction of non-crystalline material which is taken at 2θ angle of about 18°. The crystallinity index was calculated to be about 69.3% 76.5 and 78.6% for untreated fibers, bleached fibers, and cellulose nanofibrils, respectively. The increase in crystallinity of bleached jute fibers as compared to untreated jute can be explained by the removal of amorphous non-cellulosic compounds induced by the alkali and bleaching treatments performed to purify cellulose and that of jute CNF, by the removal of amorphous cellulosic domains due to acid hydrolysis.

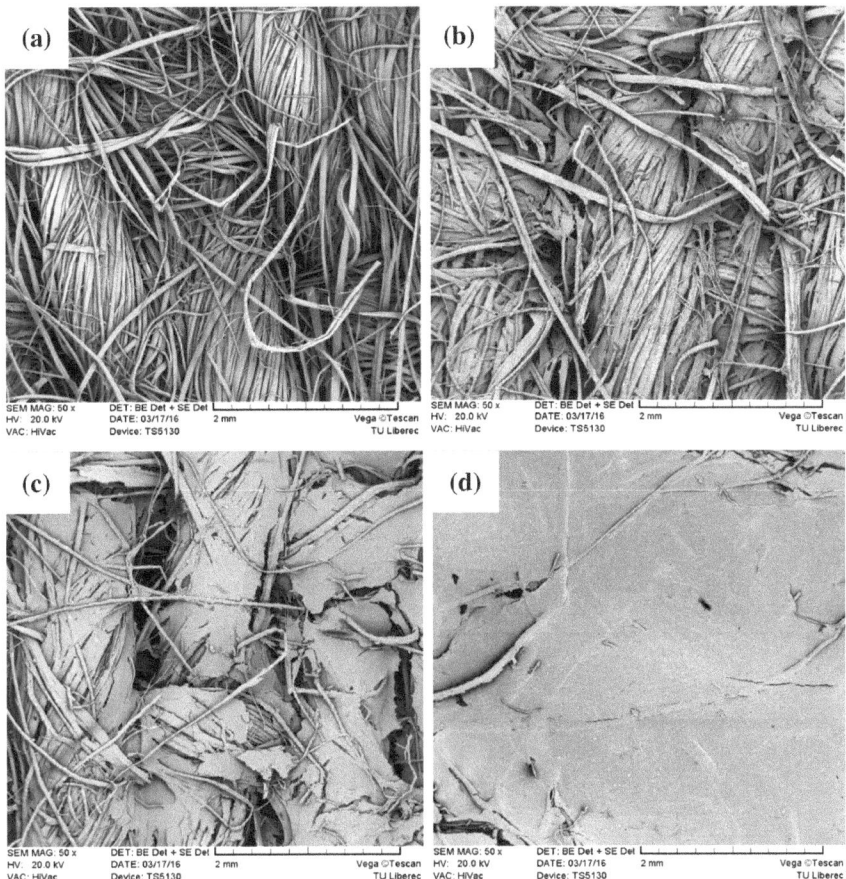

Fig. 5.3 Surface topology of jute fabric coated with; **a** 0 wt%, **b** 3 wt%, **c** 5 wt%, and **d** 10 wt% of nanocellulose suspensions

5.2.4 Tensile Properties

Figure 5.5a shows the typical stress–strain curves of uncoated and coated jute composites with different nanocellulose content. The stress–strain curve for uncoated (CF0) composite shows almost linear behavior until failure whereas the curves for composites coated with different nanocellulose content show linear behavior followed by change in slope showing nonlinear behavior, thus presenting plastic deformation and gradual debonding of fibers from the matrix just before failure. The tensile failure behavior also reveals more brittle nature of CF0 composite as compared to nanocellulose-coated composites (Fig. 5.5a).

Figure 5.5b presents the average values and standard deviation of ultimate tensile strength and tensile modulus of composites. The tensile properties indicate the decrease in strength and increase in modulus (stiffness) of composites with the

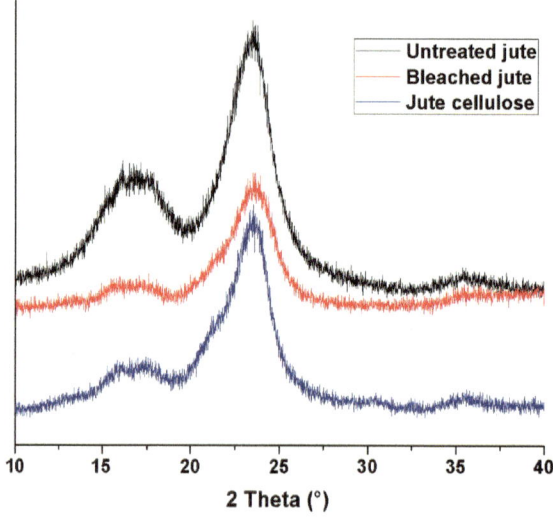

Fig. 5.4 X-ray diffraction patterns of untreated, bleached jute fibers, and jute cellulose nanofibrils

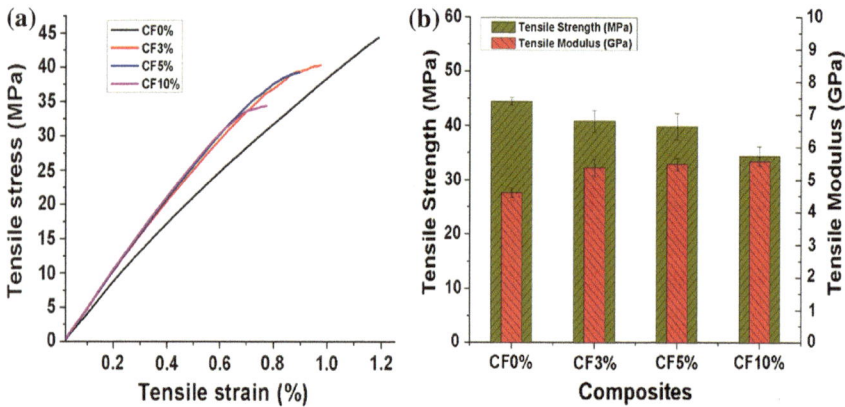

Fig. 5.5 **a** Tensile stress–strain curves and **b** tensile strength and modulus of uncoated and nanocellulose-coated jute composites

increase in nanocellulose concentration. For CF10 composite, the tensile modulus increases from 4.6 to 5.58 GPa showing 21% increase, thus presenting better fiber/matrix interfacial interaction and bonding which would be effective at the early stages of loading.

However, the decrease in tensile strength of composites with the increase in nanocellulose concentration may be explained by the differences in failure strains of nanocellulose-coated jute fabric reinforcement and the matrix. In other words, the reinforcement does not come into effect when the failure strain of matrix is much

Fig. 5.6 Flexural strength and modulus of uncoated and nanocellulose-coated jute/green epoxy composites

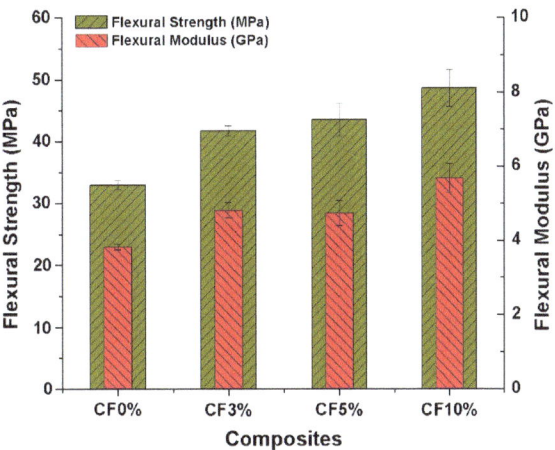

greater than that of reinforcement. So the composite shows a failure before the stress is transferred from matrix to reinforcement (Cho and Park 2011).

5.2.5   Flexural Properties

The flexural strength and modulus increase with the increase in nanocellulose concentration in composites. The flexural strength increases from 32.94 MPa for CF0 composite to 32.94, 43.53, and 48.66 MPa for CF3, CF5, and CF10 composites, respectively, thus allowing 26%, 32%, and 47% increase on average as shown in Fig. 5.6. Similarly, flexural modulus increases from 3.83 GPa for CF0 composite to 4.81, 4.73, and 5.67 GPa for CF3, CF5, and CF10 composites, respectively, thus permitting 25, 23.5, and 48% increase on average (Fig. 5.6). These findings may suggest the strong interaction between matrix and reinforcement after nanocellulose coating which increases with the increase in nanocellulose content. The other possibility of the enhancement of flexural properties may be the increase in bending stiffness/rigidity of jute reinforcement after coating with nanocellulose which also increases with the increase nanocellulose concentration (Kale et al. 2016a, b).

5.2.6   Fatigue Life

The S–N (fatigue life) curves of all composites in the considered stress range are shown in Fig. 5.7. The experimental fatigue data was fitted by semilogarithmic function, $\sigma_{max} = k \log N_f + a$ (Carvelli et al. 2016) (where k and a are the parameters to be defined by the least-squares method as given in Table 5.1), to have a

Fig. 5.7 Maximum stress (σ_{max}) versus logarithm of number of cycles to failure log(N_f) and semilogarithm fitting for each composite

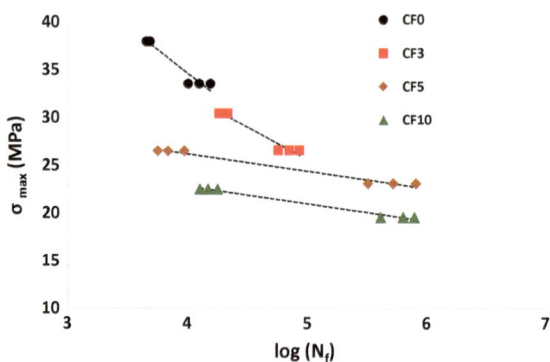

Table 5.1 Parameters of the linear fitting of S–N curves

Composites	k	a	Adj. R^2
CF0	−9.73 (3.06)	73.63 (11.90)	0.92287
CF3	−6.82 (1.69)	59.74 (7.73)	0.95104
CF5	−1.80 (0.32)	33.43 (1.55)	0.97462
CF10	−1.80 (0.26)	30.03 (1.30)	0.98312

Note Standard deviation in parentheses

reliable predicting of the fatigue life corresponding to other stress levels in the considered stress range, which are not directly determined by testing. The quality of the fitting is related to the coefficient of determination R^2 (Meloun and Militky 2011). Values of R^2 close to 1 confirm the better goodness to fit of experimental and predicted data. The curves in Fig. 5.7 have R^2 values in the range of 0.92 – 0.98 (Table 5.1). The negative slope k values of the linear fitting curves, listed in Table 5.1, show a decreasing trend with the increase in nanocellulose content of the composites. This predicts an increase in fatigue life for composites containing higher contents of nanocellulose coated over woven jute reinforcement than that of uncoated jute composite. This is clearer in Fig. 5.8 when comparing the average number of cycles to failure for the two applied stress levels. CF3 composite has higher fatigue life at 80% stress level (low cycles regime) whereas decreasing the stress level to 70% (high-cycle regime) causes a significant increase in the fatigue life of CF5 and CF10 composites.

The reliability of the above results on the fatigue life can be assessed by the confidence level index, based on the Student's t-distribution (Meloun and Militky 2011). The confidence levels in Table 5.2 show that, in strict statistics terms: the hypothesis "the nanocellulose-coated jute composites have a longer fatigue life than the uncoated jute composite (CF0)" is valid with confidence level higher than 99% for both stress levels except for CF5 composite which has confidence level >88% at 80% stress level. The better fatigue life of nanocellulose-coated jute composites can be interpreted as a better damage tolerance of these materials mainly due to increase in intermolecular and physical interactions, thus forming a rigid and stiff network due to nanocellulose coating over woven jute reinforcement.

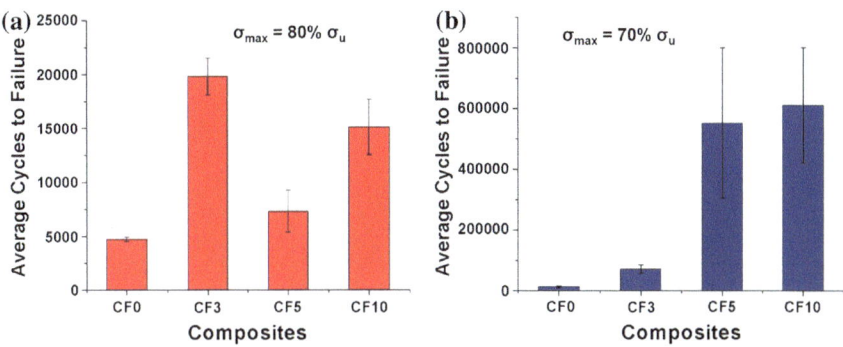

Fig. 5.8 Comparison of average fatigue life of composites: **a** 80% and **b** 70% of σ_u

Table 5.2 Confidence levels for three hypotheses (">" means longer fatigue life)

Stress level (%)	Confidence level for hypotheses		
	CF3 > CF0 (%)	CF5 > CF0 (%)	CF10 > CF0 (%)
70	99.8	99.4	99.9
80	99.9	88.8	99.3

5.2.7   Fracture Toughness

Figure 5.9a shows the typical K_Q versus displacement curves presenting the crack growth behavior of composites, and Fig. 5.9b presents the fracture toughness (K_{Ic}) with respect to nanocellulose content in composites. The fracture mode was brittle for CF0 composite showing slip-stick behavior, but as the nanocellulose content over jute reinforcement increased, the fracture mode changed from brittle to a little ductile and K_{Ic} values increased from 2.64 MPa.m$^{1/2}$ for CF0 composite to 3.20, 3.21, and 3.49 MPa.m$^{1/2}$ for CF3, CF5, and CF10 composites, respectively, resulting in 21, 21.5, and 32% increase on average (Fig. 5.9b). Thus, fracture toughness increases linearly with the increase in the concentration of nanocellulose coating over reinforcement.

Crack deflection, plastic deformation, voids, crack pinning/bridging, fiber pull-out, and debonding are the known toughening mechanisms in epoxy matrices, found in the literature (Wetzel et al. 2006). Figure 5.10a–d exhibits SEM images for all nanocellulose contents which clearly shows fiber fracture, fiber pullout, and some voids for all composites. However, fiber pullout is a little more prominent mechanism for nanocellulose-coated composites, maybe due to increased fiber debonding during fracture resulting in increased crack propagation length during deformation and hence fracture toughness.

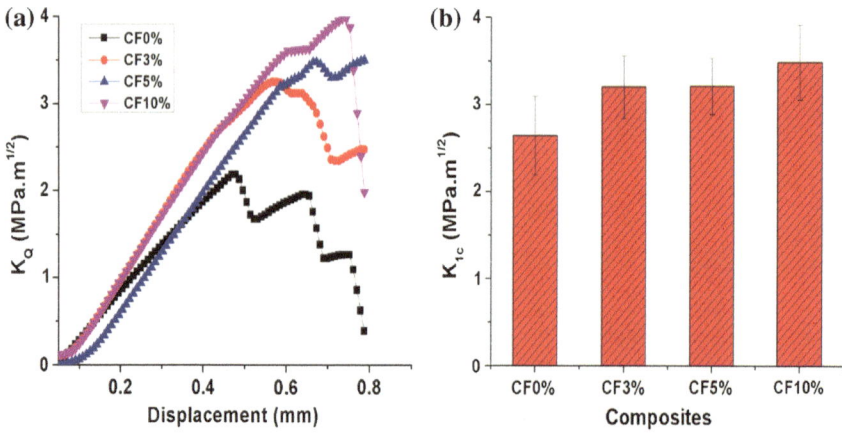

Fig. 5.9 **a** Typical K_Q versus displacement curves and **b** fracture toughness (K_{1c}) of uncoated and cellulose-coated jute/green epoxy composites

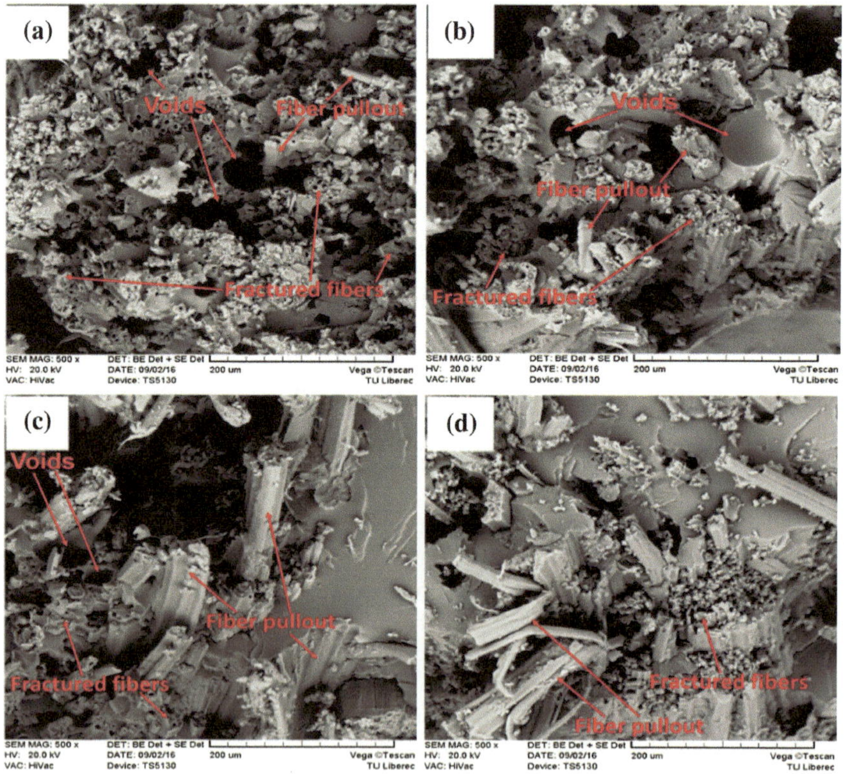

Fig. 5.10 Fracture surface topology of nanocellulose-coated jute/green epoxy composites; **a** CF0, **b** CF3, **c** CF5, and **d** CF10

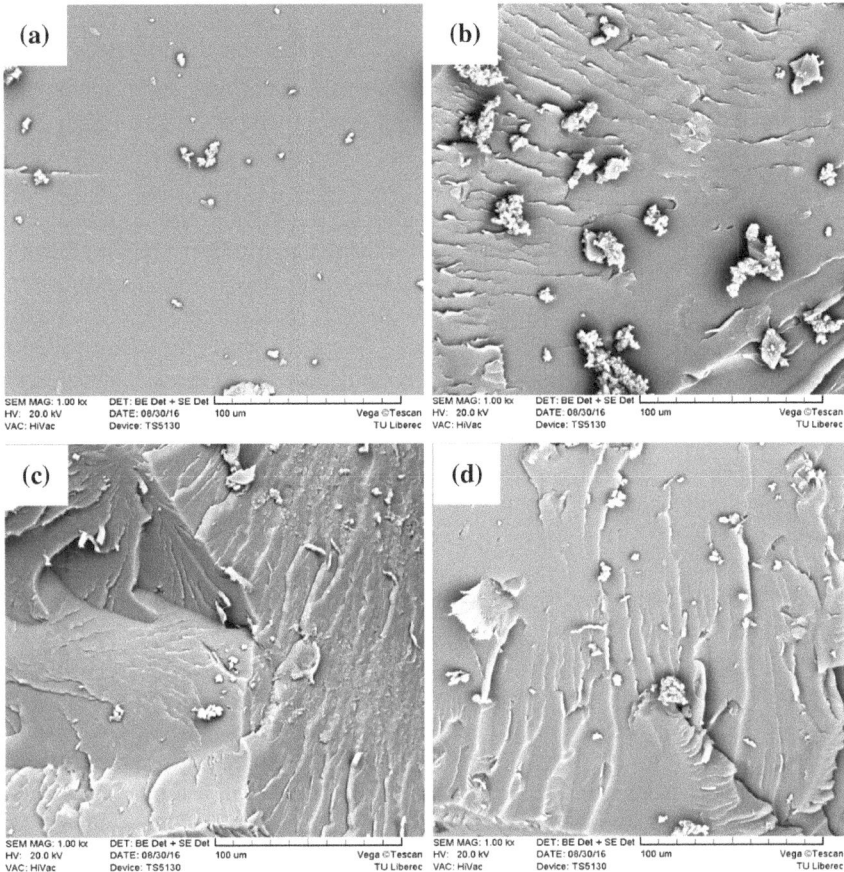

Fig. 5.11 Fracture surfaces in the matrix region of jute/green composites; **a** CF0, **b** CF3, **c** CF5, and **d** CF10

Figure 5.11 shows the fracture surface topologies in the matrix region. It is shown from Fig. 5.11a that the fracture surface of green epoxy matrix of CF0 composite is very smooth and featureless, which indicates typical brittle fracture behavior with the lack of significant toughness mechanism (Alamri and Low 2012). However, Fig. 5.11b–d shows rougher fracture surfaces and river patterns in the matrix regions of nanocellulose-coated jute composites. An increase in fracture surface roughness can be used as an indicator in the presence of plastic deformation and crack deflection mechanisms, which increase fracture toughness by increasing crack propagation length during deformation (Zhao et al. 2008).

5.2.8 Dynamic Mechanical Analysis

The change in storage modulus as a function of temperature of different nanocellulose-coated and uncoated jute composites is shown in Fig. 5.12a. It can be seen that the shape of storage modulus curves is almost same for all the samples and E' decreases with increase in temperature because of the transition from glassy to rubbery state. However, the storage modulus increases with increasing concentration of nanocellulose in composites both in glassy and rubbery regions showing superior reinforcing effects of nanocellulose-coated jute fabrics throughout the specified temperature range. In the glassy region, components are highly immobile, close and tightly packed, and intermolecular forces are strong (Pothan et al. 2003) resulting in high storage modulus, but as temperature increases, intermolecular forces become weak; the components become more mobile and lose their close-packing arrangement, resulting in loss of stiffness and hence storage modulus. Figure 5.12a reveals that uncoated jute composite (CF0) has the lowest storage modulus throughout the specified temperature range. CF0 has 4.05 GPa of E' (measured at 30 °C); however, CF3, CF5, and CF10 composites show maximum values of 4.72, 5.27, and 6.35 GPa, respectively, resulting in an increase of 16, 30, and 56% in E'. Moreover, E' (measured at 150 °C) increased from 0.28 GPa for CF0 composite to 0.48, 0.71, and 0.94 GPa for CF3, CF5, and CF10 composites, respectively, representing 71, 153, and 235% increase. This shows that when nanocellulose concentration over the jute fabric is increased, the stiffness effect of reinforcement is progressively increased not only in the glassy region but also in the rubbery region. The above findings may be attributed to the fact that, as cellulose nanofibrils coated on jute fabric possess large surface area, it promotes the interfacial interactions between the reinforcement and matrix, thus reducing the mobility of polymer chains (Chirayil et al. 2014) and better stress transfer at the interface (Ray et al. 2002). The other fact is the increase in the stiffness of reinforcement with increasing nanocellulose concentration. Furthermore, the formation of rigid and stiff network interconnected by hydrogen bonds is the accepted theory to explain the excellent mechanical properties of composites incorporated with nanocellulose (Samir et al. 2004; Lu et al. 2008; Raquez et al. 2012; Kargarzadeh et al. 2015).

Figure 5.12b presents the loss modulus (E'') versus temperature of different nanocellulose-coated and uncoated jute composites. It is observed that the value of loss modulus increased and then decreased with the increase in temperature for all composites. The rapid rise in loss modulus in a system indicates an increase in the polymer chains' free movements at higher temperatures due to a relaxation process that allows greater amounts of motion along the chains that is not possible below the glass transition temperature (Martínez-Hernández et al. 2007). Figure 5.12b also revealed that the value of loss modulus is increased with the increase in the concentration of nanocellulose coating as compared to uncoated composite (CF0) representing a higher amount of energy dissipation associated with an increase in internal friction. It is interesting to note that T_g is decreased from 95 °C for CF0 composite to 88 and 91 °C for CF3 and CF5 composites, respectively, but T_g of

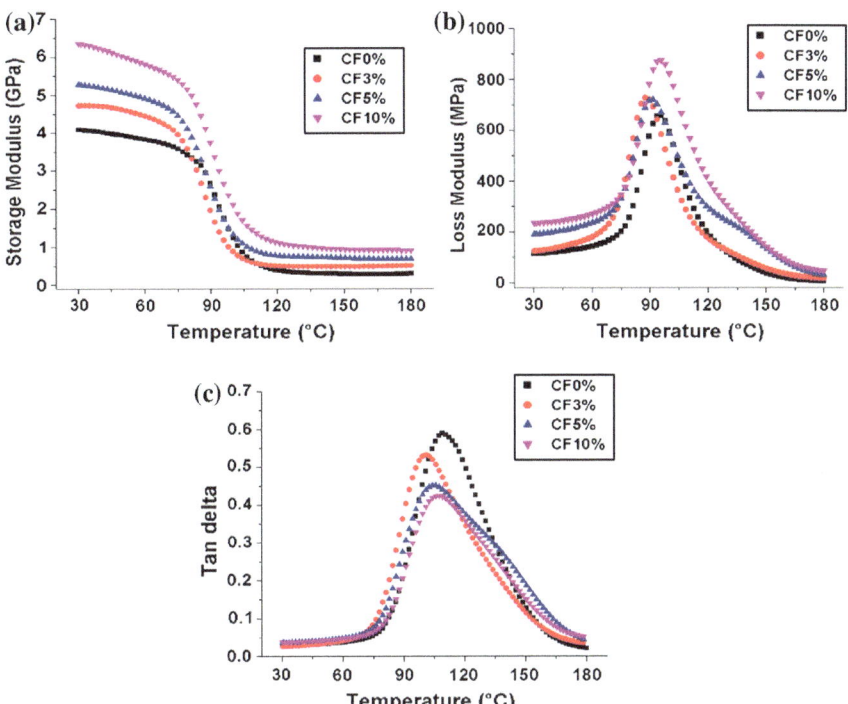

Fig. 5.12 Dynamic mechanical properties of nanocellulose-coated and uncoated jute composites; **a** storage modulus, **b** loss modulus, and **c** tan delta

CF0 and CF10 composites are almost same. Simultaneously, a positive shift in T_g values to higher temperatures is noted with the increase in nanocellulose concentration in the system (Fig. 5.12b). A possible reason may be the change in cross-linking density of the network (Brunner et al. 2006).

The damping factor (*tanδ*) as a function of temperature of composites reinforced with different concentration of nanocellulose-coated jute fabric is shown in Fig. 5.12c. The highest value of *tanδ* peak is observed for CF0 composite resulting in more energy dissipation whereas a reduction is *tanδ* peak height is observed in composites with the increase in the concentration of nanocellulose coating. The peak height of 0.59 for CF0 composite is reduced maximum to 0.42 for CF10 composite representing a 40% decrease (Fig. 5.12c), thus indicating that there are both strong intermolecular and physical interactions contributing to greater molecular restrictions at the interface and less energy dissipation and that the storage modulus is influenced more by the increase in nanocellulose concentration than loss modulus in the composites. The width of *tanδ* peak becomes broader for nanocellulose-coated jute composites, especially for CF5 and CF10. The damping in materials generally depends on the molecular motions at the interfacial region (Dong and Gauvin 1993). Therefore, the increase in width of *tanδ* peak of nanocellulose-coated composites is suggestive of increased volume of interface (Pothan et al. 2003) and an increase in the inhibition of relaxation processes in composites, thus decreasing the mobility of polymer chains within the system and a higher number of chain segments upon increase in the concentration of nanocellulose in composites (Júnior et al. 2012).

5.3 Summary

Tensile modulus, flexural strength, flexural modulus, fatigue life, and fracture toughness of composites were found to improve with the increase in the concentration of nanocellulose coating over jute reinforcement except the decrease in tensile strength. The dynamic mechanical test also revealed the increase in storage modulus and reduction in tangent delta peak height of composites with the increase in nanocellulose concentration over jute reinforcement. Based on the analysis of results, the improvement in mechanical and dynamic properties was likely attributed to the increase in interfacial interaction between reinforcement and matrix due to large surface area exposed by nanocellulose coated over jute reinforcement, the formation of rigid and stiff network interconnected by hydrogen bonds, and the increase in the stiffness of reinforcement with increasing nanocellulose concentration, whereas the differences in failure strains of coated jute reinforcement and the matrix might be the reason of reduction in tensile strength of these composites.

References

Alamri, H. & Low, I. M. (2012) Characterization of epoxy hybrid composites filled with cellulose fibers and nano-SiC, *Journal of applied polymer science, 126*(S1).

Beg, M. D. H., & Pickering, K. L. (2008). Accelerated weathering of unbleached and bleached kraft wood fibre reinforced polypropylene composites. *Polymer Degradation and Stability, 93* (10), 1939–1946.

Brunner, A., Necola, A., Rees, M., Gasser, P., Kornmann, X., Thomann, R., et al. (2006). The influence of silicate-based nano-filler on the fracture toughness of epoxy resin. *Engineering Fracture Mechanics, 73*(16), 2336–2345.

Carvelli, V., Betti, A., & Fujii, T. (2016). Fatigue and Izod impact performance of carbon plain weave textile reinforced epoxy modified with cellulose microfibrils and rubber nanoparticles. *Composites Part A: Applied Science and Manufacturing, 84,* 26–35.

Chirayil, C. J., Mathew, L., Hassan, P., Mozetic, M., & Thomas, S. (2014). Rheological behaviour of nanocellulose reinforced unsaturated polyester nanocomposites. *International Journal of Biological Macromolecules, 69,* 274–281.

Cho, M.-J., & Park, B.-D. (2011). Tensile and thermal properties of nanocellulose-reinforced poly (vinyl alcohol) nanocomposites. *Journal of Industrial and Engineering Chemistry, 17*(1), 36–40.

Dong, S., & Gauvin, R. (1993). Application of dynamic mechanical analysis for the study of the interfacial region in carbon fiber/epoxy composite materials. *Polymer Composites, 14*(5), 414–420.

Júnior, J. H. S. A., Júnior, H. L. O., Amico, S. C., & Amado, F. D. R. (2012). Study of hybrid intralaminate curaua/glass composites. *Materials and Design, 42,* 111–117.

Kale, B. M., Wiener, J., Militky, J., Rwawiire, S., Mishra, R., & Jabbar, A. (2016a). Dyeing and stiffness characteristics of cellulose-coated cotton fabric. *Cellulose, 23*(1), 981–992.

Kale, B. M., Wiener, J., Militky, J., Rwawiire, S., Mishra, R., Jacob, K. I., et al. (2016b). Coating of cellulose-TiO$_2$ nanoparticles on cotton fabric for durable photocatalytic self-cleaning and stiffness. *Carbohydrate Polymers, 150,* 107–113.

Kargarzadeh, H., Sheltami, R. M., Ahmad, I., Abdullah, I., & Dufresne, A. (2015). Cellulose nanocrystal: A promising toughening agent for unsaturated polyester nanocomposite. *Polymer, 56,* 346–357.

Lu, J., Wang, T., & Drzal, L. T. (2008). Preparation and properties of microfibrillated cellulose polyvinyl alcohol composite materials. *Composites Part A: Applied Science and Manufacturing, 39*(5), 738–746.

Martínez-Hernández, A., Velasco-Santos, C., De-Icaza, M., & Castano, V. M. (2007). Dynamical–mechanical and thermal analysis of polymeric composites reinforced with keratin biofibers from chicken feathers. *Composites Part B: Engineering, 38*(3), 405–410.

Meloun, M., & Militky, J. (2011). *Statistical data analysis: A practical guide.* Cambridge, United Kingdom: Woodhead Publishing Limited.

Mukherjee, A., Ganguly, P., & Sur, D. (1993). Structural mechanics of jute: the effects of hemicellulose or lignin removal. *Journal of the Textile Institute, 84*(3), 348–353.

Pothan, L. A., Oommen, Z., & Thomas, S. (2003). Dynamic mechanical analysis of banana fiber reinforced polyester composites. *Composites Science and Technology, 63*(2), 283–293.

Raquez, J.-M., Murena, Y., Goffin, A.-L., Habibi, Y., Ruelle, B., DeBuyl, F., et al. (2012). Surface-modification of cellulose nanowhiskers and their use as nanoreinforcers into polylactide: A sustainably-integrated approach. *Composites Science and Technology, 72*(5), 544–549.

Ray, D., Sarkar, B., Das, S., & Rana, A. (2002). Dynamic mechanical and thermal analysis of vinylester-resin-matrix composites reinforced with untreated and alkali-treated jute fibres. *Composites Science and Technology, 62*(7), 911–917.

Ray, D., Sarkar, B. K., Rana, A., & Bose, N. R. (2001). Effect of alkali treated jute fibres on composite properties. *Bulletin of Materials Science, 24*(2), 129–135.

Samir, M. A. S. A., Alloin, F., Sanchez, J.-Y., & Dufresne, A. (2004). Cellulose nanocrystals reinforced poly (oxyethylene). *Polymer, 45*(12), 4149–4157.

Terinte, N., Ibbett, R., & Schuster, K. C. (2011). Overview on native cellulose and microcrystalline cellulose I structure studied by X-ray diffraction (WAXD): Comparison between measurement techniques. *Lenzinger Berichte, 89,* 118–131.

Wetzel, B., Rosso, P., Haupert, F., & Friedrich, K. (2006). Epoxy nanocomposites–fracture and toughening mechanisms. *Engineering Fracture Mechanics, 73*(16), 2375–2398.

Zhao, S., Schadler, L. S., Hillborg, H., & Auletta, T. (2008). Improvements and mechanisms of fracture and fatigue properties of well-dispersed alumina/epoxy nanocomposites. *Composites Science and Technology, 68*(14), 2976–2982.

Chapter 6
Flexural, Creep and Dynamic Mechanical Evaluation of Novel Surface-Treated Woven Jute/Green Epoxy Composites

Abstract The focus of this chapter is to evaluate the flexural, creep, and dynamic mechanical properties of woven jute fabric-reinforced green epoxy composites as a function of modification of jute fibers by enzyme, CO_2-pulsed infrared laser, and ozone treatments. The selected treatments resulted in the enhancement of flexural properties of composites. The creep strain was experiential to increase with temperature. The treated composites exhibited less creep strain than untreated one at all temperatures. The best result in terms of creep deformation is presented by laser-treated composite which dominantly exhibited elastic behavior rather than viscous behavior, especially at higher temperatures. Dynamic mechanical analysis (DMA) results revealed that treated composites have higher storage modulus over the range of temperature. A positive shift of tangent delta peaks to higher temperature and reduction in their height for treated composites were observed. The degree of interfacial adhesion between the jute fiber and green epoxy was also anticipated using adhesion factor obtained through DMA data.

Keywords Biocomposites · Creep · Interface/interphase · Burger's model · Dynamic mechanical properties

6.1 Overview

Jute fabric is treated with some novel environment-friendly methods such as CO_2 pulsed infrared laser, ozone, enzyme, and plasma. The treated jute fibers are characterized by SEM and FTIR. The effect of novel treatments on the flexural creep and dynamic mechanical properties of prepared composites has been explored. The Burger's four parameters model is used to model the short-term creep response of composites.

© The Author(s) 2017
A. Jabbar, *Sustainable Jute-Based Composite Materials*, SpringerBriefs in Applied Sciences and Technology, DOI 10.1007/978-3-319-65457-7_6

6.2 Results and Discussion

6.2.1 SEM Observation of Jute Fibers After Surface Treatments

Significant changes in surface topology of fibers are observed after treatments. Figure 6.1a shows the multicellular nature of untreated jute fiber with a rather smooth surface, whereas a rough and fragmented surface topology can be observed for enzyme-treated fibers (Fig. 6.1b). This may be due to partial removal of cementing materials from the fiber surface after this treatment. Figure 6.1c displays the thermal degradation of surface fibers after laser treatment giving a porous and rough surface of fabric. The increase in roughness and cracks are noticeable on the surface of ozone-treated jute fiber (Fig. 6.1d). Plasma treatment causes a minor increase in fiber surface roughness. Overall, SEM micrographs give an indication that all treatments have changed the surface topology of jute fibers.

6.2.2 FTIR Analysis

The FTIR spectra of untreated and treated jute fibers are shown in Fig. 6.2. The peak at 1736 cm^{-1} is due to stretch vibration of C=O bonds in carboxylic acid and ester components of cellulose and hemicellulose and also non-conjugated carbonyls in lignin. This peak is slightly reduced for enzyme-treated fibers which show the partial removal of hemicellulose and lignin components upon treatment. However, the intensity and peak height at 1736 cm^{-1} is increased by ozone and plasma treatments. The peak at 1599 and 1508 cm^{-1} corresponds to the aromatic ring vibrations in lignin. The increase in the intensity of peak at 1736 cm^{-1} and disappearance of peak after ozone treatment at 1508 cm^{-1} is possibly due to the oxidation of lignin (Gadhe et al. 2006).

The reduction in the shoulder height at 1105 cm^{-1} and peak height at 1055 cm^{-1} for IR laser gives strong evidence that this treatment can alter the structure on the fiber surface. In addition, the increased peak intensity ~ 3200–3600 cm^{-1} for ozone- and plasma-treated fibers gives an indication of a reaction of hydroxyl bonds with the carboxyl group and reduction of peak at the same wave number range may be ascribed to a decrease of hydroxyl and carboxyl groups on the surface of laser-treated jute fiber due to thermal degradation. As a result, there is strong evidence that the used treatments have altered the surface chemistry of jute fibers.

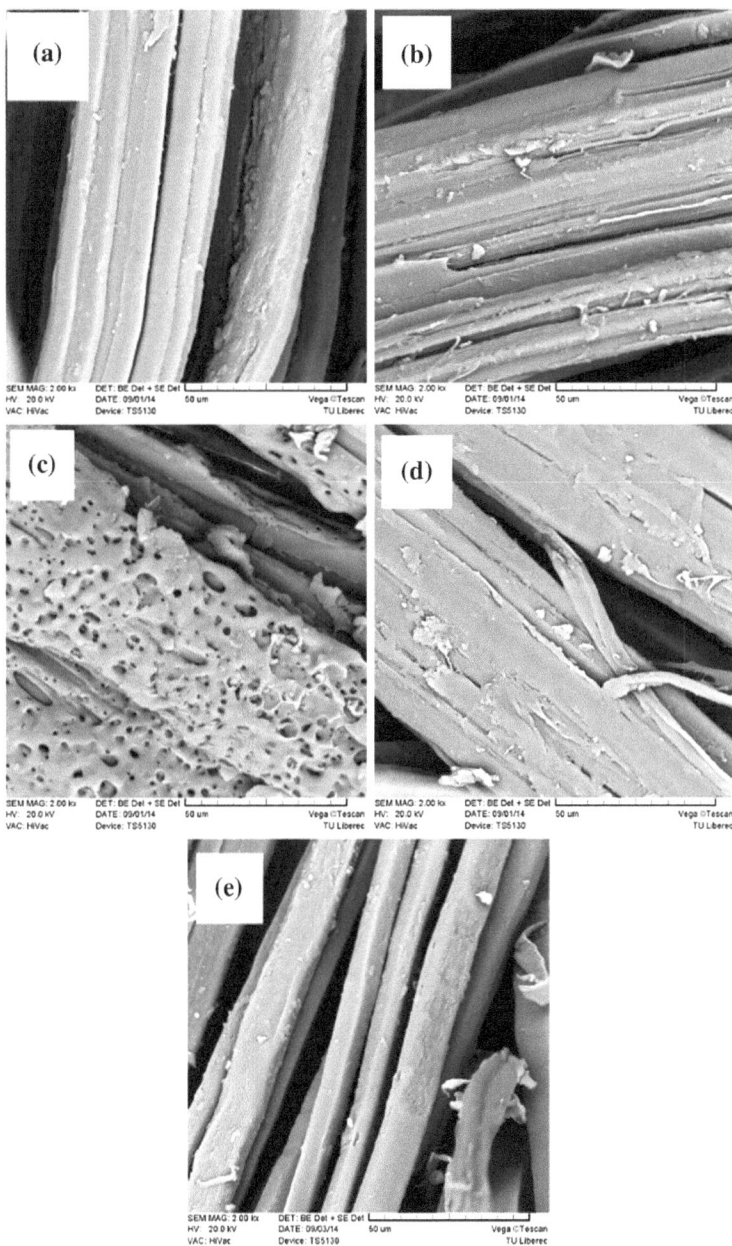

Fig. 6.1 Surface topology of jute fibers: **a** untreated, **b** enzyme, **c** laser, **d** ozone, **e** plasma

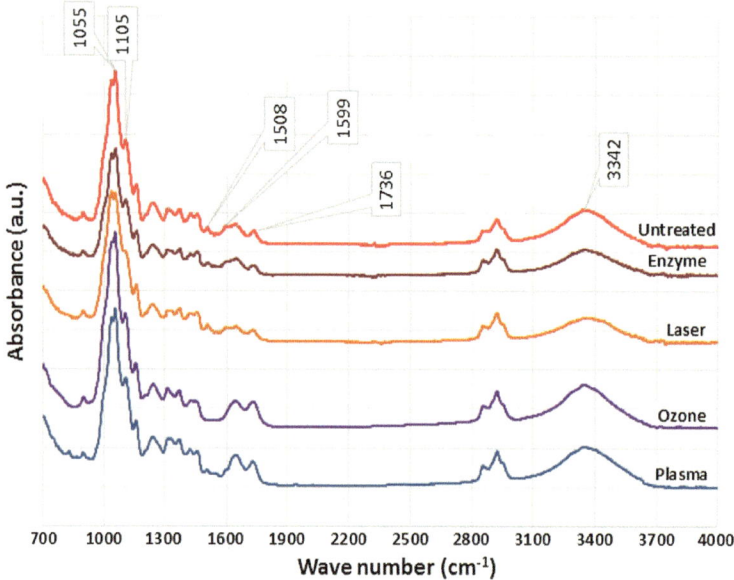

Fig. 6.2 FTIR of untreated and treated jute

6.2.3 Flexural Properties

Flexural properties of untreated and treated composites are shown in Fig. 6.3. It is interesting to note that all treated jute composites exhibit higher flexural strength and flexural modulus than untreated composite. Of the various treatments used, ozone treatment provides better flexural strength and flexural modulus, which are 13.48 and 16.16% higher than the untreated one, followed by laser treatment. The flexural strength of laser-treated composites is increased by 12.85%, and flexural modulus is increased by 13.28% compared to untreated one. It has been reported that weak fiber/matrix interfacial adhesion contributes to poor flexural properties (Khalil et al. 2007). Therefore, the possible reason for the increase in flexural properties of composites may be attributed to the increase in fiber/matrix interfacial adhesion due to treatments.

6.2.4 Creep Behavior

The creep behavior of jute composites with and without different fiber treatments at different temperatures (40, 70, and 100 °C) is shown in Fig. 6.4. The Burger's model curves show a satisfactory agreement with the experimental data (Fig. 6.4). It can be observed that the composites have low instantaneous deformation ε_M and creep strain at 40 °C due to higher stiffness of composites, but this deformation

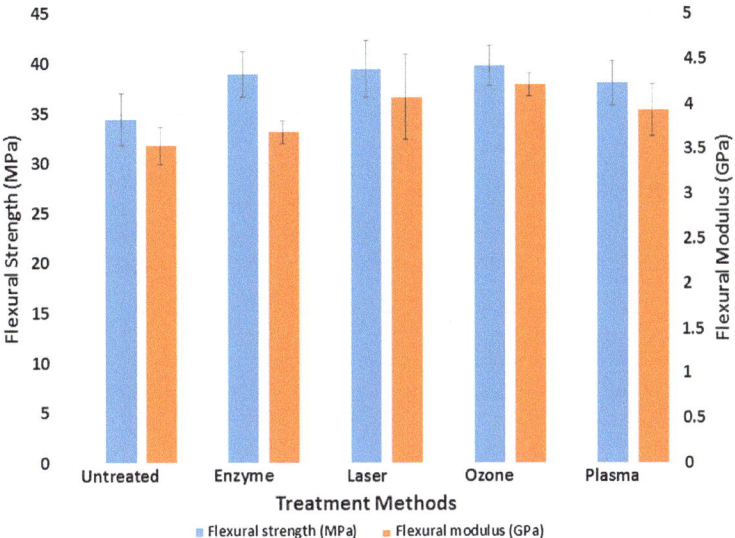

Fig. 6.3 Flexural properties of untreated and treated jute/green epoxy composites

increases at higher temperatures due to decrease in composites stiffness. The creep strain of all composites also increased at higher temperatures, but the untreated jute composite was affected more than the treated composites. When the stress is applied to the composite material, the fiber/matrix interactions are of frictional type, and shear load at the interface is responsible for the matrix/interface material flow in shear (Hidalgo-Salazar et al. 2013) and untreated composite in more prone to creep due to weak fiber/matrix interface. The four parameters E_M, E_K, η_M, η_K of Burger's model, used to fit the Eq. 2.3 to the experimental data, are summarized in Table 6.1. The first value is parameter estimator, and value in parenthesis is corresponding standard deviation. All four parameters were found to decrease for all composites as temperature increased (Table 6.1).

The parameters for untreated composites have undergone a largest decrease, resulting in higher creep strain (Fig. 6.4a). The laser- and ozone-treated composites have comparatively better values of parameters especially η_M which is related to the long-term creep strain and validates less temperature dependence of these composites.

The creep strain is low for treated jute composites at all temperatures compared to untreated one as shown in Fig. 6.5. The less creep strain is shown by laser- and ozone-treated composites at all temperatures followed by plasma- and enzyme-treated ones. The laser-treated composite has greater instantaneous elastic deformation at higher temperatures (70 and 100 °C) but less viscous deformation over time as compared to other treated composites resulting in less creep deformation. The better fiber/matrix adhesion contributes to elastic rather than viscous behavior of composite materials. The better performance of laser-treated composite may be attributed to possibility of increase in mechanical interlocking between the

Table 6.1 Summary of four parameters in Burger's model for short-term creep of the composites

Temperature	Parameters	Treatments				
		Untreated	Enzyme	Laser	Ozone	Plasma
40 °C	E_m (MPa)	2095.9 (69.4)	2685.9 (81.1)	2846.8 (88.5)	2697.7 (76.9)	2761.4 (86.4)
	E_k (MPa)	17,761.7 (7160.8)	27,557.5 (13,056.1)	38,654.8 (19,501.0)	34,269.9 (16,812.2)	43,244.2 (23,485.5)
	η_m (Pa.s)	1.67E13 (5.94E6)	2.53E13 (1.04E7)	3.57E13 (1.45E7)	4.79E13 (3.05E7)	3.44E13 (1.18E7)
	η_k (Pa.s)	1.32E6 (1.37E6)	2.28E12 (2.66E6)	1.88E12 (2.71E6)	2.28E12 (2.98E6)	1.59E12 (2.52E6)
	SS^*	4.22E-9	2.28E-9	1.53E-9	1.76E-9	1.35E-9
	Adj. R^2	0.9822	0.97795	0.96798	0.95771	0.9692
70 °C	E_m (MPa)	1673.7 (115.7)	2505.8 (90.9)	2153.9 (62.9)	2488.5 (79.2)	2976.5 (109.8)
	E_k (MPa)	3719.6 (1129.2)	11,303.05 (4056.9)	15,196.9 (5887.2)	13,682.5 (5398.7)	13,819.5 (4914.4)
	η_m (Pa.s)	5.47E12 (2.28E6)	1.31E13 (5.04E6)	1.83E13 (8.05E6)	1.85E13 (9.07E6)	1.74E13 (7.28E6)
	η_k (Pa.s)	4.36E11 (2.69E5)	1.48E12 (9.97E5)	1.68E12 (1.37E6)	1.85E12 (1.34E6)	1.71E12 (1.19E6)
	SS^*	3.71E-8	4.82E-9	3.88E-9	3.81E-9	3.43E-9
	Adj. R^2	0.98969	0.9903	0.98508	0.98696	0.98868
100 °C	E_m (MPa)	935.6 (262.6)	2225.8 (103.3)	1989.6 (63.3)	2325.5 (113.6)	2537.9 (124.0)
	E_k (MPa)	513.9 (108.7)	6710.2 (2246.4)	9710.1 (3494.3)	6157.4 (1826.2)	6202.9 (1995.9)
	η_m (Pa.s)	1.66E12 (1.05E6)	7.76E12 (2.73E6)	1.21E13 (4.96E6)	1.24E13 (6.79E6)	8.89E12 (3.64E6)
	η_k (Pa.s)	3.51E10 (1.96E4)	9.62E11 (5.62E5)	1.35E12 (8.69E5)	8.46E11 (4.53E5)	9.93E11 (5.05E5)
	SS^*	1.44E-6	1.04E-8	6.04E-9	1.04E-8	9.33E-9
	Adj. R^2	0.98555	0.99268	0.9904	0.98994	0.99314

SS^* sum of squared deviations

fiber and matrix due to formation of micropores on fiber surface (Fig. 6.1c) resulting in an increase in the shear interfacial strength and lower creep deformation of the composite.

6.2.5 Dynamic Mechanical Analysis

The variation of storage modulus (E') of untreated and treated jute fiber composites as a function of temperature at frequency of 1 Hz is shown in Fig. 6.6a. It can be seen that there is a gradual fall in the storage modulus of all treated jute composites

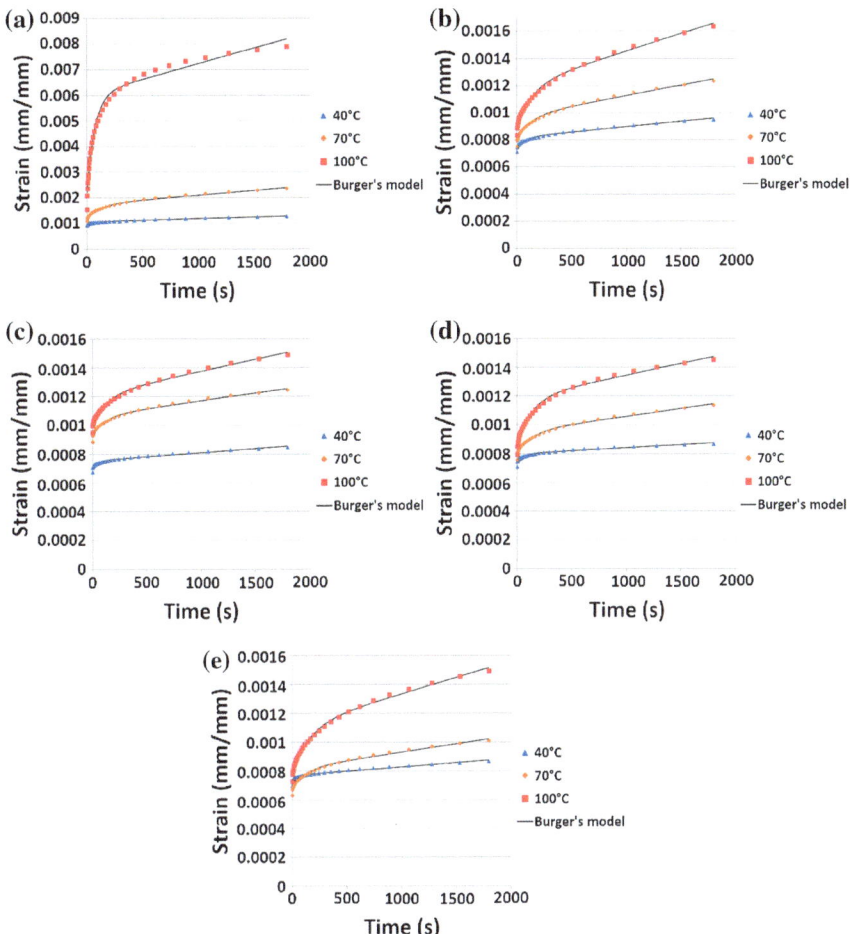

Fig. 6.4 Creep curves of untreated and surface-treated jute reinforced composites: **a** untreated, **b** enzyme, **c** laser, **d** ozone, **e** plasma

when the temperature is increased compared to untreated jute composite which had a very steep fall in E'. The DMA curves of the treated and untreated composites present two distinct regions, a glassy region and a rubbery region (Sreenivasan et al. 2015). The glassy region is below the glass transition temperature (T_g), while the rubbery region is above T_g. In the glassy region, components are highly immobile, close, and tightly packed resulting in high storage modulus (Jacob et al. 2006), but as temperature increases, the components become more mobile and lose their close-packing arrangement resulting in loss of stiffness and storage modulus. There is not a big difference in the storage modulus values of composites in the glassy region, but all treated composites have higher values of storage modulus in the

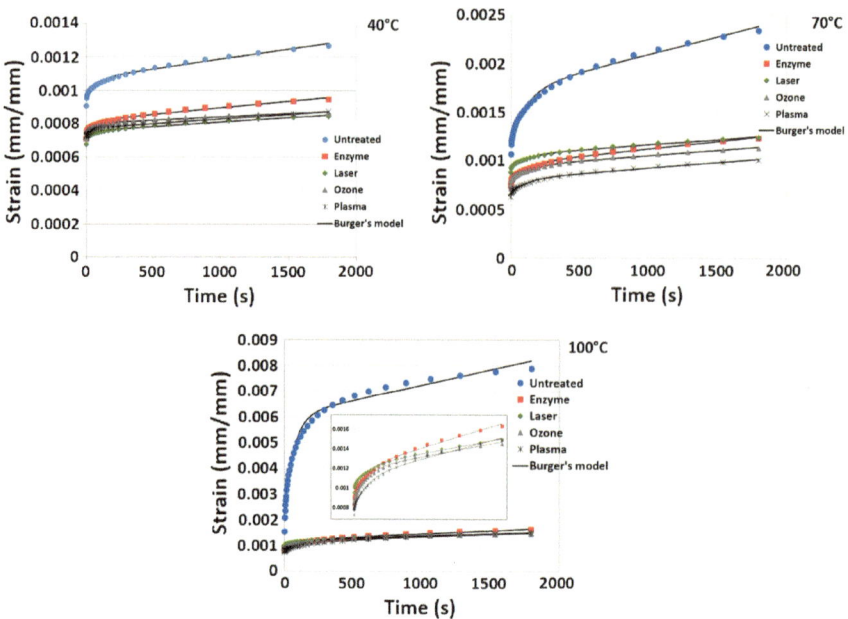

Fig. 6.5 Creep curves of composites at different temperatures

rubbery region. This might be due to better fiber/matrix interaction at the interface, decreased molecular mobility of polymer chains, and better reinforcing effect of treated fibers which increases the thermal and mechanical stability of the material at higher temperatures (John and Anandjiwala 2009) as shown prominently by laser-treated jute composite (Fig. 6.6a).

It has been reported that T_g values obtained from loss modulus (E'') are more realistic as compared to those obtained from damping factor ($tan\delta$) (Akay 1993). A positive shift in T_g to higher temperature for all treated jute composites is observed as given in Table 6.2 due to reduced mobility of matrix polymer chains and better reinforcement effect. It can be reasoned that the interfaces were markedly changed by the fiber treatments. According to Almeida et al. (Almeida Jr. et al. 2012), systems containing more restrictions and a higher degree of reinforcement tend to exhibit higher T_g. The T_g increased from 105 °C for untreated to 126–146 ° C for treated composites, especially the laser-treated one with a value of 146 °C.

The change in damping factor ($tan\delta$) of untreated and treated jute composites with respect to temperature is shown in Fig. 6.6b. Untreated composite displayed a higher $tan\delta$ peak value compared to treated composites. This may be attributed to more energy dissipation due to frictional damping at the weaker untreated fiber/matrix interface. When a composite material, consisting of fibers (essentially elastic), polymer matrix (viscoelastic), and fiber/matrix interfaces, is subjected to deformation, the deformation energy is dissipated mainly in the matrix and at the interface. If matrix, fiber volume fraction and fiber orientation are identical, as it is

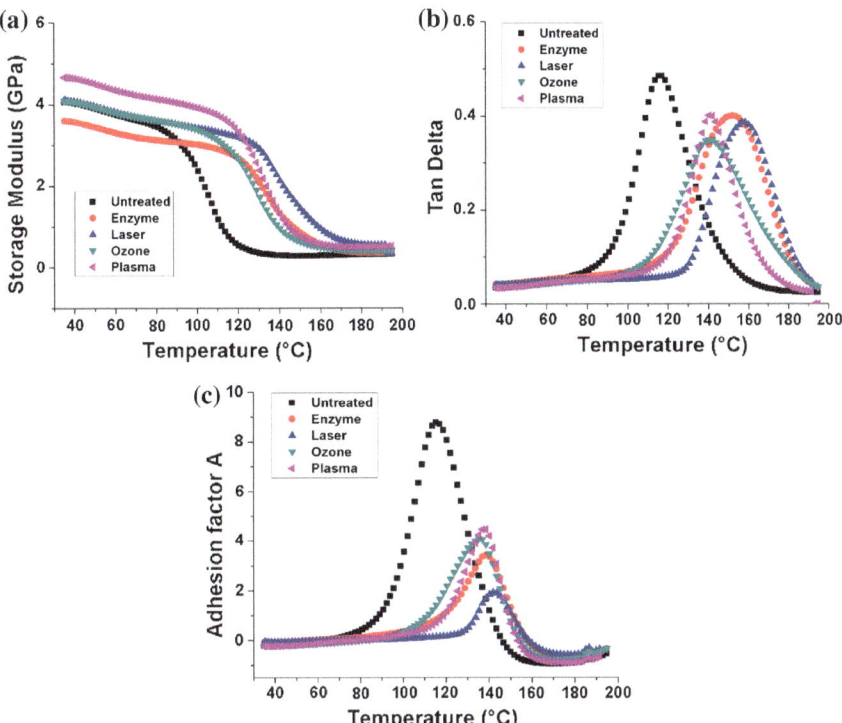

Fig. 6.6 Temperature dependence of **a** storage modulus, **b** *tan* δ, and **c** adhesion factor for untreated and treated jute composites

Table 6.2 T_g values obtained from E'' curve

Composites	T_g from E''_{max} curve (°C)
Untreated	105.60
Enzymes	137.43
Laser	146.13
Ozone	126.78
Plasma	134.57

the situation in current study, then *tanδ* can be used to evaluate the interfacial properties between fiber and matrix. The composites with poor fiber/matrix interface have a tendency to dissipate more energy than the composites with good interface bonding, i.e., poor interfacial adhesion leads to greater damping (Afaghi-Khatibi and Mai 2002; Pothan et al. 2003). The lower *tanδ* peak height is shown by ozone-treated composite followed by laser-treated one, among the treated composites, exhibiting a better adhesion between jute fibers and green epoxy matrix. The reduction in *tanδ* peak also represents the good load-bearing capacity of a particular composite (Jawaid et al. 2013). The broadening of *tanδ* peak is also observed for enzyme-, ozone-, and laser-treated samples when compared with *tanδ* peak of untreated composite (Fig. 6.6b). This indicates the occurrence of molecular relaxations at the interfacial region of composite material.

The effect of treatments on the interfacial adhesion between jute fibers and green epoxy resin was verified by adhesion factor (*A*) for fiber/matrix interface using Eq. 6.1. (Goriparthi et al. 2012);

$$A = \frac{tan\delta c(T)}{(1 - Vf)(tan\delta m(T))} - 1 \tag{6.1}$$

where *Vf* is the fiber volume fraction in the composite, $tan\delta_c$ (*T*) and $tan\delta_m$ (*T*) are the values of *tanδ* at temperature T of the composite and neat matrix, respectively. Low *A* values suggest greater interaction between the fiber and matrix. Figure 6.6c expresses low adhesion factor curves of treated fiber composites compared to untreated composite which reveals the improvement in fiber matrix adhesion with fiber treatments. Laser-treated composite has the lowest adhesion factor.

6.3 Summary

Flexural properties of composites were enhanced after fiber treatments especially for laser- and ozone-treated ones. The Burger's model fitted well the experimental creep data. The creep strain was found to increase with temperature. The treated composites showed less creep deformation than untreated one at all temperatures. The less creep deformation is shown by laser-treated composite which dominantly exhibited elastic behavior rather than viscous behavior, especially at higher temperatures. This might be due to increase in fiber/matrix interaction at the interface. Dynamic mechanical tests also established that fiber/matrix interface was modified due to fiber treatments. This is also confirmed by adhesion factor curves.

References

Afaghi-Khatibi, A., & Mai, Y.-W. (2002). Characterisation of fibre/matrix interfacial degradation under cyclic fatigue loading using dynamic mechanical analysis. *Composites Part A: Applied Science and Manufacturing, 33*(11), 1585–1592.

Akay, M. (1993). Aspects of dynamic mechanical analysis in polymeric composites. *Composites Science and Technology, 47*(4), 419–423.

Almeida, J. H. S., Jr., Ornaghi, H. L., Jr., Amico, S. C., & Amado, F. D. R. (2012). Study of hybrid intralaminate curaua/glass composites. *Materials and Design, 42,* 111–117.

Gadhe, J. B., Gupta, R. B., & Elder, T. (2006). Surface modification of lignocellulosic fibers using high-frequency ultrasound. *Cellulose, 13*(1), 9–22.

Goriparthi, B. K., Suman, K., & Rao, N. M. (2012). Effect of fiber surface treatments on mechanical and abrasive wear performance of polylactide/jute composites. *Composites Part A: Applied Science and Manufacturing, 43*(10), 1800–1808.

Hidalgo-Salazar, M. A., Mina, J. H., & Herrera-Franco, P. J. (2013). The effect of interfacial adhesion on the creep behaviour of LDPE–Al–Fique composite materials. *Composites Part B: Engineering, 55,* 345–351.

Jacob, M., Francis, B., Thomas, S., & Varughese, K. (2006). Dynamical mechanical analysis of sisal/oil palm hybrid fiber reinforced natural rubber composites. *Polymer Composites, 27*(6), 671–680.

Jawaid, M., Abdul Khalil, H., Hassan, A., Dungani, R., & Hadiyane, A. (2013). Effect of jute fibre loading on tensile and dynamic mechanical properties of oil palm epoxy composites. *Composites Part B: Engineering, 45*(1), 619–624.

John, M. J., & Anandjiwala, R. D. (2009). Chemical modification of flax reinforced polypropylene composites. *Composites Part A: Applied Science and Manufacturing, 40*(4), 442–448.

Khalil, H. A., Issam, A., Shakri, M. A., Suriani, R., & Awang, A. (2007). Conventional agro-composites from chemically modified fibres. *Industrial Crops and Products, 26*(3), 315–323.

Pothan, L. A., Oommen, Z., & Thomas, S. (2003). Dynamic mechanical analysis of banana fiber reinforced polyester composites. *Composites Science and Technology, 63*(2), 283–293.

Sreenivasan, V., Rajini, N., Alavudeen, A., & Arumugaprabu, V. (2015). Dynamic mechanical and thermo-gravimetric analysis of Sansevieria cylindrica/polyester composite: Effect of fiber length, fiber loading and chemical treatment. *Composites Part B: Engineering, 69,* 76–86.

Chapter 7
Conclusions and Future Work

Abstract The research results presented in Chaps. 4, 5 and 6 are summarized in the current chapter. The possible applications of the prepared composites are proposed, and limitations are defined. At the end, few directions for future work are suggested.

Keywords Jute/green epoxy composites · Mechanical properties · Dynamic mechanical properties · Creep models · Automotive applications

In this research work, woven jute/green epoxy composites were prepared using three different categories of reinforcement viz. pulverized micro jute fillers, nanocellulose-coated jute fabrics, and novel surface-treated jute fabrics. Mechanical, dynamic mechanical, and creep properties were evaluated. The modeling of short-term creep data was satisfactorily conducted using Burger's model. The long-term creep performance of PJF-filled jute composites was successfully predicted by using Burger's model, Findley's power law model, and a simpler two-parameter power law model. The following findings were drawn from the results.

7.1 Mechanical Properties

1. The tensile and flexural properties were found to improve with the incorporation of PJF in alkali-treated jute/green epoxy composites except the decrease in tensile strength of composite reinforced with only alkali-treated jute fabric.
2. Tensile modulus, flexural strength, flexural modulus, fatigue life, and fracture toughness of composites were found to improve with the increase in concentration of nanocellulose coating over jute reinforcement except the decrease in tensile strength.
3. Flexural properties of composites were enhanced after fiber treatments especially for laser- and ozone-treated ones.

7.2   Creep Behavior

1. The creep resistance of PJF-filled jute composites was found to improve significantly with the increase of filler content in matrix. This may be attributed to
the inhibited mobility of polymer matrix molecular chains initiated by large
interfacial contact area of PJF as well as their interfacial interaction with the
polymer matrix.
2. The surface-treated jute composites showed less creep deformation than
untreated one at all temperatures. The least creep deformation is shown by
laser-treated composite which dominantly exhibited elastic behavior rather than
viscous behavior, especially at higher temperatures.
3. The Burger's model fitted well the short-term creep data. The creep strain was
found to increase with temperature.
4. The master curves, generated by time–temperature superposition principle
(TTSP), indicated the prediction of the long-term performance of composites.
5. The Findley's power law model was satisfactory for fitting and predicting the
long-term creep performance of composites compared to Burger and
two-parameter model.

7.3   Dynamic Mechanical Properties

1. Dynamic mechanical analysis revealed the increase in storage modulus and
reduction in tangent delta peak height of composites with the increase in filler
content and increase in concentration of nanocellulose coating over jute reinforcement and of surface-treated jute composites.

7.4   Proposed Applications and Limitations

The possible applications of these composites can be in automotive interiors
especially, boot liners, door panels, spare tire cover, and interior vehicle linings.
Despite the advantages of sustainability of these composites and cheaper availability of jute fibers, the main drawback is a little high cost of green epoxy resin
compared to synthetic ones. The utilization of jute waste as precursor to extract and
purify cellulose is a positive aspect of this work. The methods used in this study for
fiber treatment especially, the ozone and laser, are environment-friendly and easy to
perform, but the laser treatment can be applied effectively on a little thick fabric
substrate.

7.5   Future Work

This work has endeavored to introduce jute waste as a cheaper and easily available source to extract cellulose and also novel environment-friendly methods for surface modification of natural fibers. Due to scope of the task, following recommendations are suggested for future work.

1. Investigation of aging and durability of composites due to weathering effect in order to have a better assessment of product life cycle.
2. Analysis and modeling of the recovery behavior of composites after creep deformation.
3. Study of the fire properties of composites by addition of suitable additives.
4. Study of the impact properties of composites especially by drop weight impact testing.